Modeling Technoscience and Nanotechnology Assessment

# COMPARATIVE STUDIES ON EDUCATION, CULTURE AND TECHNOLOGY
# VERGLEICHENDE STUDIEN ZUR BILDUNG, KULTUR UND TECHNIK

Edited by / Herausgegeben von Tomasz Stępień

Advisory Committee
Rev. Prof. Jerzy Machnacz (Pontifical Faculty of Theology Wrocław, Poland)
Prof. Jan Misiewicz (Wrocław University of Technology, Poland)
Dr. Annette Deschner (University of Education at Karlsruhe, Germany)
Prof. Alicja Kalus (University of Opole, Poland)
Dr. Mojca Kompara (University of Primorska, Slovenia)
Prof. Henryk Kiereś (Catholic University of Lublin, Poland)

## VOLUME 4

*Zu Qualitätssicherung und Peer Review der vorliegenden Publikation*

Die Qualität der in dieser Reihe erscheinenden Arbeiten wird vor der Publikation durch den Herausgeber der Reihe geprüft.

*Notes on the quality assurance and peer review of this publication*

Prior to publication, the quality of the work published in this series is reviewed by the editor of the series.

Ewa Bińczyk / Tomasz Stępień

# Modeling Technoscience and Nanotechnology Assessment
Perspectives and Dilemmas

**Bibliographic Information published by the Deutsche Nationalbibliothek**
The Deutsche Nationalbibliothek lists this publication in the Deutsche Nationalbibliografie; detailed bibliographic data is available in the internet at http://dnb.d-nb.de.

**Library of Congress Cataloging-in-Publication Data**
Modeling technoscience and nanotechnology assessment : perspectives and dilemmas / Binczyk, Ewa [and] Stepien, Tomasz.
   pages cm. — (Comparative studies on education, culture and technology ; volume 4 = Vergleichende Studien zur Bildung, Kultur und Technik ; Band 4)
   Includes bibliographical references and index.
   ISBN 978-3-631-64735-6 (print) — ISBN 978-3-653-04371-6 (E-book) 1. Technology—Philosophy. 2. Technology—Social aspects—Simulation methods. I. Binczyk, Ewa, 1976- author. II. Stepien, Tomasz, 1968- author.
   T14.M566 2014
   303.48'3—dc23
                                        2014027194

This publication was financially supported by
the Wrocław University of Technology.

ISSN 2196-5129
ISBN 978-3-631-64735-6 (Print)
E-ISBN 978-3-653-04371-6 (E-Book)
DOI 10.3726/978-3-653-04371-6
© Peter Lang GmbH
Internationaler Verlag der Wissenschaften
Frankfurt am Main 2014
All rights reserved.
Peter Lang Edition is an Imprint of Peter Lang GmbH.

Peter Lang – Frankfurt am Main · Bern · Bruxelles · New York ·
Oxford · Warszawa · Wien

All parts of this publication are protected by copyright.
Any utilisation outside the strict limits of the copyright law, without the permission of the publisher, is forbidden and liable to prosecution. This applies in particular to reproductions, translations, microfilming, and storage and processing in electronic retrieval systems

This publication has been peer reviewed.

www.peterlang.com

# Table of content

Foreword ..................................................................................................... vii

Caring about the future of the collective.
Monitoring technoscience in the sociology of risk
and science and technology studies (by Ewa Bińczyk) ............................. 1

A. Starting point ......................................................................................... 3
The modern shape of technoscience ......................................................... 6

B. Proposals ............................................................................................... 15
Beck's cosmopolitanism and the challenges posed by the risk society ............... 16
Latour's "politics of nature" ...................................................................... 23
Transformation of the boundary conditions
of the public debate on innovation ........................................................... 32
Concrete ways of expanding participation (and their
respective weaknesses) .............................................................................. 42
The Precautionary Principle, Technology Assessment (TA)
and select examples of other institutional solutions ............................... 48
The macro-ethics of global responsibility ................................................ 55
Conclusion ................................................................................................ 62
Literature ................................................................................................... 67

Nanotechnology. Assessment and Convergence inside
the Technoscience (by Tomasz Stępień) ................................................... 77

A. Constitution of Nano-Domain as Science and Technology ................ 77
1. The background of technological convergence: From
'normal' to 'post-normal' science ............................................................. 77
   1.1. Theoretical framework of nanotechnology ...................................... 78
   1.2. Methodological aspects of nanotechnology ..................................... 87
   1.3. Dimensions of nanotechnology and the necessity
      of modified theory of science ............................................................. 94
2. Science and society relationship in the case of nano-domain ............. 99
   2.1. Nanotechnology and the concept of a socially robust science ........ 100

2.2. Political background and societal impacts of nano-domain ............... 103
2.3. Nanotechnology and the public opinion: Politics of nano-images ..... 112

B. Nanotechnology and Development of Assessment Regime ........................ 118
3. General issues of risk assessment by converging technologies ................... 118
4. Nanotechnology: risk assessment and precautionary principle ................. 122
   4.1. Characteristics of risk in nanotechnology ............................................. 125
   4.2. Risk management in nanotechnology .................................................... 128
   4.3. Strategies by nano-risk communication and
        nano-visions assessment ........................................................................ 134

C. Ethical Aspects: Nano-Safety and Nanotoxicology ........................................ 142
5. Nano-Safety and Nanotoxicology ................................................................... 144
6. Ethical, legal and societal implications of nanotechnology ........................ 150
7. Nanoethics or explorative philosophy of nanotechnology ......................... 160

D. Convergence and Nanotechnology Assessment ............................................. 166
8. Converging technologies: Categories, principles and fields ....................... 166
9. Principle of technological convergence ......................................................... 173
10. Multidisciplinary framework of nanotechnology assessment ..................... 183

References ................................................................................................................ 193

Index ........................................................................................................................ 203

# Foreword

We present the book that was jointly prepared as a result of cooperation between two authors. The first part of the book, entitled "Caring about the future of the collective. Monitoring technoscience in the sociology of risk and science and technology studies" was written by Ewa Bińczyk – a philosopher of science from Nicolaus Copernicus University. Bińczyk discusses various postulates of a theoretical, social and political nature that have been formulated in response to the problem of the unwanted side-effects of the practical success of technoscience. They derive from two theoretical perspectives: the contemporary study of risk and science and technology studies (STS), inspired by what is called the actor-network theory (ANT).

She starts by asking about the very nature of modern science, commercialized and politicized in many aspects. In her part of the book Bińczyk builds the space for normative and institutional reactions to the problem of risk. We find there analyses of political and philosophical propositions, macro-ethical postulates, as well as concrete proposals for legal solutions and projects in the field of social policy and management. All of them are systematically presented, critically commented on and, in the last part of the book, reformulated in coherent a way as possible.

Bińczyk convinces us that we are faced with a particular conceptual revolution, taking place in front of our eyes. It establishes a new vocabulary for our thinking about politics, technology, responsibility and the civilizational role of technoscience. This conceptual revolution challenges our routinely accepted beliefs on society, the dynamics of social change, effectiveness, laboratories, infrastructures, progress and the need for economic growth.

Tomasz Stępień – a philosopher of culture and technology from Wrocław University of Technology – in the second part of the book with the title "Nanotechnology. Assessment and Convergence inside Technoscience" convincingly illustrates the postulates and general theses formulated by Bińczyk. He concentrates upon the very details of nanoscience and nanotechnology. His account derives from a familiar but slightly different tradition, namely: technology assessment. He analyses and characterizes nano-domain as an example of the development of techno-sciences from the point of view of philosophy of technology including three main theoretical approaches: technology assessment (TA), science and technology studies (STS) and converging technologies (CT). The starting point in his part of the book is the theoretical and methodological constitution of nano-domain as science and technology, what makes out the framework of nanotechnology assessment regime with questions related with the nanotoxicology and the Precautionary Principle. In the following Stępień presents the ongoing debate with controversies and dilemmas in the field of nano-ethics. The last part is an endeavour to design a multidisciplinary framework of technology assessment with nanotechnology as condition of technological convergence. Generally, in the case of nanotechnology it is an attempt to calibrate reciprocally to each other the indeed familiar but also slightly different theoretical approaches established in the present philosophy of science and technology.

We easily find numerous common assumptions characterizing both narrations. First of all, both authors jointly understand science as 'post-academic', 'post-normal' and socially robust. Both parts of the book picture university as entrepreneurial (as they write: knowing and producing are never separated). Taking these circumstances into account both Bińczyk and Stępień agree that the principle of self-regulation of science and research must be radically questioned. In fact, they articulate the need for *a new contract* between science and society (instead of the false linear model of autonomous science). As they emphasize, technoscience in general and nanoscience/nanotechnology in particular must remain transparent to the public and subjected to moral/political responsibility.

The book convinces us that the diffuse distribution of responsibility in the contemporary risk society is one of our most urgent political concerns. Yet, Bińczyk and Stępień postulate a theoretical attitude towards technoscience in general and nanotechnology in particular that is situated beyond utopian promises or dystopian fears. They notice that the problem of equity (e.g. fair distribution of profits and benefits of a given technology) must be stated openly, according to the rules of inter-generational distributive justice. As they underline, in fact each technology increases existing inequalities of good distribution.

Both philosophers emphasize the strategic role of the humanities in the development of key technologies. The humanities provide the necessary reflecting attitude. They can play an important role in moderating dialogue, monitoring laboratories, creating common languages and vocabularies, implementing public discussion and identifying possible obstacles to it. At the same time, they critically highlight the fact that one public with homogeneous and predictable views does not exist. While creating public dialogues about future innovations we need to take it into account.

In the face of the problems of delayed or unforeseeable side-effects of discoveries and innovations both Bińczyk and Stępień discuss the very reasonableness and possible ways of implementing the Precautionary Principle. They underline the need to avoid the Collingridge dilemma related to technology assessment (that assessment is always implemented too soon or too late). Stępień emphasizes that too much speculation too soon may result in the creation of unwanted artificial public concern. According to this author, what is especially important is that nanotechnology assessment is exactly 'at the right time'. In this domain 'ethics first' is currently implemented in front of our eyes – a reflection about possible, unwanted, social, ethical or political side-effects is here implemented as an integral part on the basic-research level. Therefore, in the domain of nanotechnology we have a good chance to observe how these processes of monitoring work. Both authors hope that creating successful nanotechnology assessment may constitute for us a good opportunity of learning how to create successful public debates also about other innovations and discoveries.

Ewa Bińczyk, Tomasz Stępień

Ewa Bińczyk

# Caring about the future of the collective. Monitoring technoscience in the sociology of risk and science and technology studies[1]

> *We must most of all lower our arrogance decibels.*
> Immanuel Wallerstein, *The End of the World As We Know It*

When identifying the major political challenges facing the risk society, Ulrich Beck invokes the metaphor of a plane journey – he claims that we are asked to embark upon a journey whose destination remains uncertain and unknown. He writes: "the actors who are supposed to be the guarantors of security and rationality – the state, science and industry – are engaged in a highly ambivalent game. They are also no longer trustees but suspects, no longer managers of risks but also sources of risks. For they are urging the population to climb into an aircraft for which a landing strip has not yet been built" (Beck 2008: 12). Among these "suspect" agencies, the German sociologist lists the state, science and industry. These are not trustworthy as areas that can guarantee the stability of the course.

Various postulates of a theoretical, social and political nature have been formulated in response to the problem of the unintended consequences of the practical success of science and technology. Our goal will be to present and critically discuss proposals of this type, which derive in large measure from the study of risk and science and technology studies (STS). We shall seek to define the space for normative and institutional reactions to the problem of risk. We shall analyse political and philosophical reflections, chosen conceptual redefinitions, ethical postulates, as well as concrete proposals for legal solutions and projects in the field of social policy and management. What form do they currently take? In which direction should they be heading?

---

[1] This is a translation of a reformulated, previously published fragment of my book, *Technonauka w społeczeństwie ryzyka. Filozofia wobec niepożądanych następstw praktycznego sukcesu nauki* [Technoscience in a risk society. Philosophy on the unintended consequences of the practical success of science] (Wydawnictwo Naukowe UMK, Toruń, 2012, p. 271–382). In the text, I use the notion of "collective" (instead of "society"), introduced by Bruno Latour in order to underline the interdependency of people and non-human actors. I shall expound upon this conceptual decision below.

We will start by asking about the nature of modern science, in particular its ties to industry and the regulations imposed by the state. A presentation of previously neglected determinants of technoscience, situated within an ever tightening network of market and political forces, will be the starting point for further reflection on normative, ethical and political topics.

We shall also strive to understand the reasons for the situation described by Beck. This shall take numerous paths. For we will attempt to identify the underpinnings of the peculiar helplessness of the main agencies of modern society when it comes to the problem of the unintended consequences of the practical success of science. Is it possible to indicate the mechanisms that legitimise and support the said helplessness? In what ways are they connected to the assumptions shaping the public's view of experts, science and technology? How does our view of society and the dynamics of social change influence our perception of the problem of risk? Finally, what types of mental prejudgements allow us to ignore risk or redefine this problem as insignificant?

We know that modern risk derives from the intensification of global, heterogeneous connections. It arises from the impacts of technoscience and industry. From a certain moment in history, scientific laboratories have enabled the professional, effective and intensive mobilisation of an uncountable multitude of new actors (facts, artefacts, technologies, micro-organisms, medical solutions, environmental problems). Entering into unexpected relationships with the entities already peopling the collective, these constructs have started to modify the latter's nature, often resulting in destabilisation.

I would like to underline that the arguments presented below do not lead to the conclusion that the further "evolution" of humanity, or the "advance" of technologies, should be conservatively blocked. We have no other choice but to accept that to a certain extent, every novelty or change has a certain potential for destabilisation, or may even be a threat. Literature and the humanities as well as the ideas they introduce also bring about destabilisation. The phenomenon is not uniquely proper to technoscience. We can reach the same conclusion regarding mass media or the internet, whose effects are instantaneous and have a global reach. Technoscience should therefore not be blocked. Rather, the point is to have a public discussion about its true role and link to market mechanisms.

The authors of the positions outlined below mostly express their opposition to the arrogance of current solutions, which demonstrate a tendency to disregard the problem of risk. They urge us to modify our ingrained ways of thinking about the role of science and technology, to reflect upon the values that determine our political choices and to create institutions and resolutions of a systemic, institutionalised nature. I shall present these postulates and discuss, or in certain cases, criticise them. I will address the question of the risk generated by technoscience (and our responsibility to deal with it) in a two ways: globally, and to a lesser extent, individually, while also, in certain cases, referring to the modification of individual attitudes. In the last part of my argument I shall present the conclusions that, in my opinion, are the most important.

## A. Starting point

Finding an answer to the question stated above requires, above all, taking into account the real dependencies obtaining within the socio-economic structures in which technoscience (a key area when it comes to generating innovation and discoveries that have destabilising potential) is currently entangled. Taking technoscience out of its political context, with which it is deeply interwoven, seems harmful in the long run. To believe in the absolute innocence and independence of scientific research is pure wishful thinking which blocks any examination of the true role of science in global society. It does not allow us to see the real consequences of our own actions. This type of thinking seems inadequate. As Andrzej Szahaj, a Polish political philosopher, notes: "The paradigm of the axiological neutrality of science shows too many anomalies" (Szahaj 2007: 160).

Modern science is not autonomous; it is commercialised, post-academic and driven by military needs.[2] Furthermore, it is to be expected that the commercialisation of science will only increase (Mirowski, Van Horn 2005: 537; Jones 2009: 843). The global economic order resting on market mechanisms is without doubt the most important framework in which the consequences of the practical success of technoscience as well as its current post-academic form should be examined.

---

2 Science has always been enmeshed in various types of dependence. Considering it autonomous is a purely rhetorical statement. Representatives of the scientific community have also always been involved in the publicising of knowledge, as well as in legitimising and promoting their own activities. They have also often been politically active. Clear delimitations between disciplines did not exist once. What is interesting, British and US industry in the 18th and 19th centuries encouraged scientists to concentrate on fundamental (and not applied) research, keeping the autonomous role of universities in high regard. The European universities of the 19th century were above all tasked with educating elites, not conducting research. They did not do applied research at all. However, scientists have always acted as consultants for the mining, steel or tool industries (Godin 1998: 468–469, 473).

Economic globalisation has mainly brought about the abolition of trade barriers, the introduction of currency exchange rates, the lowering of transport and communication costs, and more generally, the lowering of transaction costs as well as increased market predictability (at least for the most important market players). According to Joseph E. Stiglitz, winner of the 2001 Nobel Prize in economics, globalisation is the "closer integration of the countries and peoples of the world which has been brought about by the enormous reduction of costs of transportation and communication, and the breaking down of artificial barriers to the flows of goods, services, capital, knowledge, and (to a lesser extent) people across borders" (Stiglitz 2002: 2002).

On the other hand, Jagdish Bhagwati writes that globalisation (in the economic sense) consists in processes of market integration, which exist thanks to international trade, direct foreign investment, short-term capital flows, workforce mobility and technology transfers (Bhagwati 2004: 3). In the opinion of this economist, the processes of economic globalisation, taken in this sense, were set in motion by the lifting of trade barriers in the 19th century.

Unfortunately, due to the thematic choices underpinning the current argument, the dependence of technoscience on the global and market contexts cannot be discussed here in detail. Let us therefore merely highlight a few chosen issues. Above all, it should be emphasised that the mechanisms underlying globalisation not only shorten distances and facilitate the revolution in communications described above, but also lead to the immediate transfer of risks, crises and threats, conflicts, group panics or epidemics. For globalisation unites "too rapidly and without negotiation" (Latour 2002: 24).

In particular, the interests of the weakest, marginalised groups, and even of whole nations, are not negotiated. Furthermore, there is no strong, proper articulation of the interests of future generations – though ecological movements speak in their name – that voice is quite often disregarded. We also still lack the tools or institutions that would ensure the protection and proper representation of the interests of non-human actors: the atmosphere, endangered species, ecosystems. As we can see, the most forceful processes within globalisation in the economic sense are often implemented by way of *faits accomplis*, murkily, without a democratic debate in which the collective could weigh alternative values against the hegemony of market drives.

Obviously, popular beliefs as to the benefits of the so-called free market (one of the key hopes of the project of modernity) are based on a broader foundation of philosophical assumptions and axiological preferences, which are not always stated directly. According to these, people are, and always will be naturally governed by self-interest, profit and the maximisation of consumption, at the same time disposing of a faculty enabling them to rationally identify these drives. The existence of these natural tendencies and needs in turn justifies increased investment and production as well as the promotion of competition within the sphere of free trade. The free market, which guarantees a framework for the implementation of the above ideals, is considered the healthiest context for the evolution of humanity. It is seen to contain self-healing mechanisms, while marketization is considered to naturally improve economic systems and democratise societies, being beneficial to the promotion of human rights. Precisely because of this, the consolidation of local markets and economies into a worldwide system, known as economic globalisation, is supposed to benefit all participants. The beliefs presented here co-exist with a trust in scientific and technical "progress", which are seen to lead to increased prosperity.

As it turns out, the condition of modern societies, in particular with regard to the problem of risk, but also the dramatic rise in social inequality (which I unfortunately cannot address here)[3] force us to seriously question the suppositions outlined above on which the global economic order is based. As Immanuel Wallerstein writes, it is not science and technology, but capitalism that is the core issue (Wallerstein 2004: 115). The free market provides the channels and mechanisms by which scientific discoveries and technological innovations spread within collectives, encountering neither difficulties nor barriers. It is also the market that fuels the constant need for novelty, driving innovation. We must remember that it is precisely within the context of the normative free market preferences that the questions of risk and the possible monitoring of technoscience are posed. Without questioning the ideal of consumption growth, we often want *both* more goods produced, as well as more trees (Wallerstein 2004: 110).

Many researchers express their disillusionment with the capitalist system, accusing it of not fulfilling its promises. It has deepened socio-economic disparities between people and has proven destructive to the biosphere. The disregard of producers when it comes to environmental damage is part of the logic of capitalism. It is profitable, it lowers costs. Entrepreneurs always strive to transfer the costs of production "outside", elsewhere, by "externalising" them. This can mean to a country, an employee, a consumer, but also to the environment. But what has

---

3   We can confidently state that the concentration of the world's wealth has currently reached neo-feudal proportions. From 1960 to 2000, the share of the richest 20% in the world's income has risen from 70 to 90%, while the share of the poorest has fallen from 2.3 to around 1% (Beck 2005: 51). According to a 2006 UN report, the "richest 2% of the world's adult population possesses more than half of the world's wealth" (Klein 2008: 490).

now become the pressing problem of the modern capitalist system is that there is nowhere left to externalise these costs. For instance, there is no rural population that could serve as a low-paid workforce in industry. Wallerstein writes: "We must require all production organizations to internalize all costs, including all costs necessary to ensure that their productive activity neither pollutes nor uses up the resources of the biosphere" – the costs of repairing damage must be included in the costs of production (Wallerstein 1998: 80).

The French sociologist Bruno Latour draws a similar picture of the modern situation. In his view, we find ourselves at a point where there is no longer an exterior to human *praxis*. There are no remaining areas to which we can externalise the costs of our conduct. There is nowhere we can dump the unwanted consequences of our activities, be it toxic waste or toxic debts (Latour 2009: 8). Beck shares this view: environmental problems can no longer be externalised outside of the collective; they are at the centre of our institutional order (Beck 2000: 224).

For humanity, which had placed successive parts of its environment within the realm of its actions, what was "outside" existed only as a potentiality. Currently, that external environment has been drastically reduced. The post-environmentalist paradox is that the concept of "environment" has been introduced into public debate at a time where a pristine environment no longer exists. The term "natural environment" suggested the existence of a certain external context to our activities, while in the current situation we can no longer speak of anything of the sort. It is hard to speak today of a natural environment uncontaminated by human actions, since it is shaped by our ethical and political choices.

Humankind will soon have to confront a situation in which human interference in the environment, economy and culture, due to the extent of the interconnections between various areas of collective life, will come back to haunt us, often with increased force. Because of this, we should be cautious in introducing further innovations, and devise systemic solutions which guarantee the stability of the collective.

## The modern shape of technoscience

> *What do you think MIT, where I work, is? It is an enormous public corporation where technologies are developed for the needs of private companies at the expense of the taxpayer.*
>
> Noam Chomsky

Asking, in one of the chapters of his book *Science in Action*, about *who* precisely makes discoveries and introduces innovations, Latour defines technoscience as an intrinsically collective endeavour. Science ceases to be effective when it isolates itself from the wider context. The more it is interconnected with it, the better for research. The list of actors we should mention here is very long, in particular if we take into account the role of mobilising allies and resources, legitimising one's own positions, seeking recruits, or, finally, the importance of persuading the public as to the importance of research and the correctness of research results. Within technoscience, many actors are involved in the effort to popularise discoveries, formulating arguments aimed at various contexts: political, economic, related to social policy, or religious. Indeed, differentiating between an "outside" context and "purely" theoretical research "inside" science does not seem particularly useful. Keeping this in mind, let us however try to consider the phenomenon of technoscience in the context of the most characteristic dependencies defining this domain.

In his book, Latour cites statistical data concerning spending on science in the United States as well as a number of western countries in the first half of the 1980s. Let us start with these statistics. And so, in 1982 it was reported that R&D (*Research and Development*)[4] consumed 2.6% of the gross national product (GNP) of the United States, 2.2% in England, 2.6% in France, 2% in Germany, 2.4% in Japan and 3.6% in the USSR (Latour 1987: 168, 162–173). What is interesting, the United States had only spent 0.2% of its GDP on research and development in 1929, and already as much as 3% in the 1960s (Bucchi 2004: 7–8).[5]

On the other hand, according to the 2007 figures published by the OECD countries (*Main Science and Technology Indicators*), those countries spend, on average, 2.29% of their gross national product (GNP) on R&D. For Israel, this indicator is 4.68%, 3.47% for Japan, 3.47% for South Korea, 2.68% for the United States, 2.54% for Germany, 2.08% for France and 1.12% for Russia. On average, European Union countries spent 1.77% of their GNP on R&D in 2007, while Poland spent 0.57%.[6] Therefore, most western countries spend around 2–3% of their GNP on research and development.

As we can imagine, the most important technoscientific resources are concentrated in a small number of leading institutions. In this respect, developing countries remain dramatically behind. We can observe a concentration of elites in science. Here, success drives further success, while a prestigious intellectual

---

4   Research and development is most often understood as long-term, future-orientated, group projects. It provides knowledge that guarantees the introduction of new solutions for the technological industry.
5   The increase of research spending in the west in the 1960s is explained by the so-called Sputnik effect – the need to compete with the USSR in terms of innovation after the Soviet Union placed the first artificial satellite in orbit in 1957.
6   Cf. www.nsf.gov/statistics/seind10/pdf/c04/pdf, accessed 9.10.2011.

environment stimulates the researchers themselves, allowing them to develop with increased intensity. It is estimated that 50% of all research papers are published by 10% of all authors, while 80% of citations pertain to 20% of the most quoted works (Sismondo 2010: 36). The United States are the centre of scientific development in many fields, attracting the top researchers. According to a ranking published by *Newsweek International,* the 2006 list of the best universities in the world contained 51 institutions in the United States, 32 in Europe, 10 in Asia and 7 in Australia (Mucha 2009: 12). There are around 35 000 scientific journals worldwide, 7 000 of which are published in the United States (Krimsky 2006: 244). As Latour provocatively states: "Half of technoscience is an American business" (Latour 1987: 167).

However, it is worth noting that the US monopoly on scientific development has been challenged. In the early 1990s, Japan surpassed the USA in terms of pure numbers of active scientists and engineers, becoming the world's leading nation in this respect (Bucchi 2004: 11). Impressive changes are underway in Chinese and Indian science, and Europe is a leader when it comes to technical universities.

Only a quarter of researchers and engineers conduct research and development work, which is the tip of the technoscience iceberg. Three quarters of engineers and scientists are "those who help in the definition, negotiation, management, regulation, inspection, teaching, sale, repair, belief and spread of the facts" (Latour 1987: 164). These (significant) actors went unnoticed in the traditional reflection on science, the rationalising philosophy of science or classical sociology of knowledge.

Modern science is characterised by interdisciplinarity, the involvement of a variety of entities and institutions in research, as well as strong pressure to use scientific discoveries in the short term (Bucchi 2004: 134). Given the growing requirement for the rapid commercialisation of scientific discoveries, the postulate that science has entered a post-academic phase is mooted ever more often. John Ziman, a physicist and philosopher of science, introduced the notion of post-academic science in his work *Real Science* (Ziman 2000). He writes of science that is: 1) academic 2) industrial (ruled by market forces) and 3) post-academic (where the distinction between academic and industrial science is lost). In his view, in its post academic phase the activity of scholars is not only connected with the political and economic dimension, but also with the ethical. It therefore cannot be examined outside of the context of social responsibility.

Writing about the fluid barrier between fundamental and applied research, Latour estimates that pure research currently constitutes no more than 10% of what we call technoscience: "Of nine dollars spent, only one goes for what is classically called 'science'. Technoscience is on the whole a matter of development"

(Latour 1987: 169). What is more, for 1982, we can estimate that around 75% of research and development was *carried out* within industry (Latour 1987: 170). As we can read in another paper, fundamental research accounted for around 17% of the total spending on science in the United States in 2004, 2006 and 2008 (Afeltowicz 2011: 12).

Significant changes took place in the 1980s in the relationship between universities and industry in the United States. Of particular importance were legislative changes leading to increased competitiveness and commercialisation. The breakthrough was the Bayh-Dole Act of 1980, which allowed university researchers to patent inventions developed with federal government funding (Bayh-Dole Act 1980; Slaughter, Rhoades 2002; Sismondo 2010: 192). Before the Bayh-Dole Act, this happened only in exceptional circumstances and required a complicated procedure. The Act commoditised knowledge, opening an era of academic capitalism. Scientific discoveries became products, licensed and sold. Patent offices were opened on campuses and partnership programmes with industry were launched. Continuous education channels were established, students were hired by industry and technologies were licensed, bought by industry for universities.

This was the beginning of the privatisation of education, which took on an ever more interdisciplinary character, becoming more diversified in terms of organisation, adjusting for a fast market use of its achievements. A high level of commercialisation has been particularly visible during the last few years in the dynamically developing field of the biomedical sciences.[7] The link between scientists and industry is particularly strong in high technology areas. For instance, 31% of biotechnologists in the US are caught up in these ties (Godin 1998: 474).

An analysis of the process of the commercialisation of science from the 1980s onwards, at least in the key US biopharmaceutical sector, requires taking into account not only the changes at universities, but also the considerable systemic role of Contract Research Organizations (CROs). These non-academic, non-public

---

7   Mark Peter Jones (Jones 2009) writes on the beginning of the commercialisation of biomedical research resulting in conflicts of interest, presenting the example of Ivor Roystor – an immunologist at the University of California, San Diego, who became a multimillionaire within seven years. Still in the 1970s, the transfer of biological research material was not subject to formal property regulations (cell lines could be gifted, transferred to private companies and so forth). The process of their commercialisation did not meet with wider resistance from researchers. The study of Roystor's career shows how the rapid transformation of the biomedical sciences destabilised the legal system (Jones 2009: 831).

organisations practically did not exist before. They have completely transformed the rules for the functioning of laboratories, and make incredible profits. They offer research services relating to the invention and introduction of pharmaceuticals, as well as their clinical trials to pharmaceutical companies (Mirowski, Van Horn 2005: 505 and following). Conflicts of interest are not a problem for these organisations, since they are by definition driven by profit and operate quickly and effectively.

Other legal regulations also have played a significant role in the process of commercialisation. In 1982, despite opposition from public institutions, the Small Business Innovation Development Act allowed small companies to apply for federal research grants. In turn, the Federal Technology Transfer Act of 1986 allowed federal laboratories to sign partnership agreements with the private sector and universities. The role of universities in these legislative changes was minimal (Slaughter, Rhoades 2002: 86–89). However, their consequences proved significant: a tremendous increase in the incomes of scientists engaged in partnerships with industry, structural changes within universities, a change in the working methods of laboratories, a wider gulf between science and the humanities. These transformations have contributed incrementally to the declining significance of fundamental research. And it would seem that few entities or social groups are interested in fighting to reclaim autonomy for US universities (Slaughter, Rhoades 2002: 103).

Currently, the patent system is also becoming more integrated and standardised, which has important economic and institutional consequences (for instance for the direction of scientific research). These are changes that endanger the traditional ethics of "pure" science. As we know, living organisms are also patentable in the US. In 1980, the American Supreme Court allowed the first strain of bacteria, *Pseudomonas aeruginosa*, to be patented. In 1988, the first animal, *Oncomouse TM*, was patented. In 2001, a patent was awarded for human embryo haemocytoblasts, as well as embryos obtained by parthenogenesis from unfertilised mammalian eggs. A further step was the patenting of genes, including human ones. In June 2000, the Human Genome Sciences Company was the owner of around 100 human gene patents (Krimsky 2006: 99). Fees for the use of patented human DNA sequences raise the costs of research, including research on pharmaceutical drugs and diseases. According to a survey carried out in Belgium, as many as 60% of scientists left research projects due to intellectual property concerns (Lave, Mirowski, Randalls 2010). In the field of biotechnology, patentability even applies to elements of the research process itself, including research tools used in biomedicine such as cell receptors (Mirowski, Van Horn 2005: 523–525). As it therefore turns out, patenting blocks the exchange of scientific knowledge and ends the era of the unselfish spread of scientific discoveries, at least in certain domains.

According to Robert Merton, an American sociologist of science, the success and authority of science derive from its norms. In his essay *Science and Democratic Social Structures* we can read that the ethos of science is constituted by the following principles: organised scepticism, disinterestedness, universalism (which claims that the criteria for judging scientific discoveries should be based on merit) and communalism (a rule that presupposes communal, public ownership of scientific discoveries) (Merton 2002: 582–583).[8]

However, the commercialisation of scientific achievements undermines the norms of universalism, disinterestedness and communalism. Often, it even makes it impossible to support them institutionally. Consequently, scientists act more like entrepreneurs, while venture capitalism establishes research programmes (Bucchi 2004: 134). This is particularly true of domains at the cutting edge of progress, such as nanotechnology, biotechnology, microelectronics, telecommunications (Mucha 2009: 107). Today's science, financed by industry, the military or by governments is more reminiscent of a market ruled by rival consultants (Yearley 2005: xii). The term "technoscience", which sees ever broader use, aptly reflects the tendency towards a decline in the significance of fundamental research and the blending of science with the implementation of its results.

Scientific research is also increasingly more often financed from non-public funds and performed outside of public universities. As Benoît Godin notes, during the 1990s, around 5–10% of research and development was financed by industry in OECD countries. However, this data does include individual consulting work. In 1995, Godin conducted his own research in Quebec on the levels of individual involvement of researchers in consulting or contracting work for industry or government. He analysed responses from over 1500 individuals (Godin 1998: 475). Out of the scientists he questioned, 44.4% did consulting work, while 28.5% were involved in the transfer of knowledge or the commercialisation of research. As it happened, among those doing consulting work, 40% collaborated with industry, while 60% collaborated with public institutions. The most scientifically productive researchers were also those with the highest number of extra-academic connections (Godin 1998: 477).

---

8   It is worth noting that Merton's model is currently seen as more wishful than empirical (Sismondo 2010: 27), an effect of his incorrect methodological decisions – i.e. basing the model on a sampling of statements made by prominent scientists and philosophers (Fuller 2006: 15).

Currently, in countries such as Germany or Sweden, the business sector finances between 10–15% of the public research budget, while in Japan, 73.4% of research is financed by industry (Bucchi 2004: 135). It seems that this tendency is on the rise. Bucchi writes: "It is calculated that around 64 per cent of research world-wide is financed by companies and that almost 70 per cent of it is performed by the companies themselves" (Bucchi 2004: 135).

Observing these changes, Rebecca Lave, Philip Mirowski and Samuel Randalls openly cite the influence of neo-liberal policy[9] on western science (mostly US) starting in the 1970s. They note the following tendencies: 1) a clear decrease in public funds allocated to universities, 2) a separation between teaching and research, the former becoming less about comprehensive education and more about producing human capital, 3) increased percentages of temporary employment in science,[10] 4) the dismantling of the institution of authorship, the spread of phenomena such as ghost authorship and ghost management, as when leading scientists append their names to already completed texts and reports, 5) the limiting of research programmes to subjects dictated by commercial needs, 6) the commodification of intellectual property, blocking universal access to knowledge (Lave, Mirowski, Randalls 2010: 659, 667; cf. Mirowski, Van Horn 2005: 527 and following). The commercialisation of science is not limited to biomedicine and biotechnology but also includes other sciences (meteorology, environmental sciences, botany, marine biology), and even the social sciences and law. These trends are decidedly different from what was called the commercialisation of science in the 1960s and 1970s. As I have underlined, the transformation of patent law alone impacts not only the organisation of science, but also its methods and content.

Due to the privatisation and commercialisation of science, it is hard to maintain a clear division between fundamental and applied research. As Gernot Böhme notes: "the famous unity of research and science has come apart, the most important and interesting research is done elsewhere – in extra-university research institutions and industry" (Böhme 1998: 14). Fundamental research is therefore quite poorly funded as compared to research with mostly practical and technological aims, financed by global corporations or the military. It is not states, but consortia that finance the most significant technical and scientific

---

9  In the understanding of these authors, neo-liberal policy assumes an expansion of market relationships into the public sphere, as well as the active building of the conditions for free trade through a strong state and transnational institutions such as the World Bank, IMF or WTO.
10 In 2005, over 48% of scientific posts at US public universities were temporary or partial (Lave, Mirowski, Randalls 2010: 665).

projects (Beck 2005: 185). The experts employed by these consortia also prepare national and transnational agreements on environmental protection, investment and financial markets.

Within science and technology studies, a distinction is made between moderately autonomous sciences, fundamental in nature and developed as part of particular disciplines (Mode-1 science) and the interdisciplinary "production of knowledge", presupposing the fastest possible implementation (Mode-2 knowledge production) (Gibbons et al. 1994; Nowotny, Scott, Gibbons 2001). The knowledge production system today is also subject to significant, and probably irreversible, changes. This new manner of producing knowledge gradually came into existence after the Second World War, partly because of the new mass character of higher education and even the massification of research work. Another reason is the intensification of international industrial competition in a global era (Gibbons et al. 1994: 70–89, 47). This causes purely theoretical research to disappear, since it is too expensive within the sciences and unnecessary within the humanities at a time where literature is available over the internet. The production of scientific knowledge (also within the humanities) is performed in an interdisciplinary context and conditioned by heterogenous dependencies (social and economic). Instead of academic interests and hierarchy, values related to the context of application predominate. In the most crucial scientific domains, we can observe transitional organisational structures and not permanent institutions. Scientific research is performed within state agencies, industry lobbies, think-tanks, research centres, consulting firms and international corporations. This is accompanied by a growing awareness of economic conditions and one's responsibility towards those who commission the research, which results in a weakening of the overall autonomy of science. Research results are not communicated through traditional channels, such as conferences and journals, which prevent the accumulation of scientific discoveries. However, they can be reused within surprising contexts, thanks to the strategic links or electronic ties between heterogeneous actors. This particularly concerns areas such as the biomedical sciences, environmental research (clean technologies), material science, nanotechnology and superconductors. Past notions of "applied science" or even "research and development" are no longer adequate to describe these phenomena, while a fundamental distinction between "pure" and applied science, industry and universities, loses its meaning.

Current interdisciplinary research is not driven by scientific curiosity but strongly geared towards solving the specific practical problems indicated by those who finance it. Its theoretical and methodological core is often only locally instantiated and stabilised. Research groups are transitory, and after completing work on a given project the experts do not return to their former disciplines. Science has many sources of funding today (Gibbons et al. 1994: 29, 78-79).

Science today is increasingly permeated by economic considerations as well as an awareness of its dependence on public needs. Scientific research and ethics committees find it problematic to properly instil reflection in science. The traditional organisation of science by discipline does not allow for the implementation of mechanisms that would sensitise one to the problems of public consequences and responsibility. Newer, transdisciplinary forms of knowledge production, where the issue of sensitivity to social and economic consequences is immediately raised by lawyers and businessmen taking part in the projects in question, are better suited to this task. The demands for science, concerning the need to include the views of the users of a given innovation and to recognise the importance of ethical and environmental issues, are rising. Due to an increase in the number of criteria, success in the second type of science (Mode-2 knowledge production) is also defined differently. The predominance of the transdisciplinary model of knowledge production and of the globalisation of information has led to a concentration of skills in the hands of a small group of top researchers, and thereby to increasingly unequal access to knowledge.

Quite obviously, technoscience, like all other areas of culture today, is subject to acceleration, computerisation and globalisation. Thanks to new communication technologies, scientists are able to collect, catalogue, exchange and manipulate data with increasing speed. It is also worth emphasising the non-negligible influence of mass media on science. Research on the public image of science shows that the press most often presents scientific activity as progressive and socially beneficial. However, the media have a tendency to place greater emphasis on certain areas (e.g. biomedicine instead of mathematics), while also exaggerating the risks associated with the progress of technoscience (Bucchi 2004: 109-110). This defines the public image of experts and orientates the discussion on innovation.

The influence of the media on science is not limited to popularisation. The media themselves shape scientific theories, while also deciding their credibility. The mass media have contributed to the spread of certain metaphors (e.g. that of the Big Bang) and have also accelerated the forming of a consensus on the theory of a hole in the ozone layer (Bucchi 2004: 118). When we examine the functioning of science today within the context of the effects of mass media, we find that it is hard to maintain a distinction between a discovery and its popularisation. The last few years have also seen intensified efforts to popularise scientific achievements (through books, television programmes and even specialised channels that enjoy high audience ratings).

To conclude, let us present the contemporary characteristics of technoscience as described in the studies cited above:

1) the systemic, institutionalised dismantling of academic science and its ethos, decreased significance of universities,
2) concentration of elites,
3) the decline of the role of fundamental research, in particular within domains crucial to the future of the collective,
4) the blurring of the boundary between "pure" and applied science, the birth of a new type of knowledge production: transdisciplinary, heterogeneous, containing transitional organisational structures, aimed at the quick implementation of discoveries,
5) an increase in the level of non-public funding for research, more research performed outside of public institutions,
6) an intensive commercialisation of research from the 1980s onwards (legislative changes, the role of CROs, exemption from state oversight),
7) consolidation of the US patent system, commodification of intellectual property.

## B. Proposals

Let us now outline the most valuable reactions to the problem of the unintended consequences of the success of science and technology. Each of the proposals for monitoring technoscience (and in a wider sense, the future of the collective) outlined below shall be treated as a more or less consistent project, based on a rather scattered literature. For the comfort of the reader, we shall attempt to present a summary of each of proposal. The final part of the text will be concerned with general conclusions.

Within some of the positions outlined below (such as those of Beck or Anthony Giddens), the problem of modern risk is equated with that of the most visible environmental hazards (mostly climate change). However, it is worth keeping in mind that the unintended consequences of discoveries made in laboratories can manifest themselves in very diverse fields: law, morality, social ties, and even the collective imagination.

Bearing in mind this important caveat, let us begin with Beck's quite general position. Examining the details of this project, I shall cite a few select theses by important authors (mostly the interesting position of a Polish sociologist, Andrzej Zybertowicz) formulating analogous political postulates within the context of the challenges posed by the era of risk.

## Beck's cosmopolitanism and the challenges posed by the risk society

In the opinion of Beck, the originator of the theory of the risk society, science is *systemically blind* to the problem of its unintended consequences.[11] It is unable to perceive the problem of risk, the nature of global connections and the strong and often constitutive link between non-humans and the social world. Modern science is much too specialised to grasp the unintended consequences of its own success. It only combats the symptoms of threats or the most blatant effects of mankind's actions, mainly through "environmental protection policies", which *de facto* boil down to temporary solutions and evasion tactics.

For instance, Beck shows that the notion of limit values, encountered in the scientific debate on ecological risks, serves an ideological purpose. By setting an acceptable limit for contamination, pollution or toxicity, we legitimise the *status quo*, minimising risk only in the short term. This is reminiscent of superficial response policies, which avoid confronting the actual problems of modernity head on. By accepting limit values, we act short-sightedly.

Our thinking about the role of technoscience is still influenced by modernist beliefs: in the unquestioned value of freedom of research, freedom of investment and of technical and scientific progress. In the public sphere, science also demands the maintenance of its hitherto recognised monopoly on rationality, linked to the special epistemological role of experts. However, this monopoly is starting to disappear before our very eyes (Beck 2002: 255).

As it follows from Beck's analyses, science as an institution noticeably avoids redefining its deepest assumptions, in particular the axiomatic assumption that science itself constitutes an unproblematic good, while the discovery of facts and the continuation of research (without concerns as to its subject, scope or goals) is valuable *a priori*. The belief that science delivers knowledge as an unproblematic good, and that its evolution will only have worthwhile consequences for humankind is one of the main tenets of scientism. It is in the attitude of scientism that the ideal of modern science is expressed. According to this ideal, the activity of scientists (regardless of all else) is morally unquestionable, since they are striving for Truth. As I also suppose, this belief in the intrinsic "innocence" of science and laboratories is also legitimised by essentialist assumptions. Since the essence of reality is given and unchanging, science only serves to discover and reveal it. This cannot be ethically problematic, since the discovery of Truth is always good. Within the essentialist framework, one loses sight of the fact that laboratory work also consists of effective actions, manipulations, interventions and the

---

11   The same position is advanced by Niklas Luhmann (Luhmann 1991: 207).

introduction into the collective domain of previously unknown and wondrous possibilities.

Zybertowicz strongly questions these scientistic assumptions present in our thinking. He postulates: "rejecting the assumption that knowledge – provided it is verifiable, intersubjective etc. – is an unproblematic good, should become an element of the ethics of science" (Zybertowicz 2003: 101). By introducing numerous innovations through market channels, technoscience fragments the social world. Processes of desacralisation and marketization in particular facilitate the introduction of changes.[12] According to Zybertowicz, an important condition for the easy and successful introduction of innovations is that the process does not require a remodelling of the existing social fabric, values, patterns of thinking, practices and ways of coordination. Technological innovations, as opposed to axiological turning points or changes within the symbolic layer, usually do not require complex cultural reconfiguration. They can more easily colonise domains outside of the laboratory. It is easier to introduce mobile telephony than equality between men and women. Modern science, based on market mechanisms, creates a "constant stream of innovations and scientific legitimations", generating chaos within cultures. In effect, modern society ceases to be transparent to itself. Social and ethical norms remain behind new structures of action.

As a remedy to the situation, Zybertowicz stresses the need to reveal the links between science and capital, and calls for slowing down the development of knowledge. To achieve this: 1) science should be desacralised by demonstrating that it (co-)creates reality instead of uncovering it, 2) it should be demonstrated that science is not a privileged domain for solving the problems plaguing humankind, 3) mechanisms should be devised for the effective monitoring and limitation of science (Zybertowicz 2003a: 70). However, due to the current rate of changes, this will not be easy to implement. "The ceaseless evolution of technology, stimulated by scientific discoveries, makes social institutions (including state ones) so dynamic as to exclude good organisation and policy", the scholar observes pessimistically (Zybertowicz 2003a: 70).

One of Beck's main theses is that modern risk requires urgent political solutions, precisely because it is a serious, long-term systemic problem. First of all, we require an awareness of the existence of modern risk. As Beck notes, risk

---

12 By providing a naturalistic, sociological explanation for the roots of the success of the natural sciences, Zybertowicz concentrates on how the *ideas* produced by sciences interact with wider social structures and in what way the *beliefs* produced by them find niches for their expansion (Zybertowicz 1999: 14, 27). This view is therefore (too strongly, to my mind) concentrated on the intellectual and belief layer.

increases when we refuse to accept its existence (as in the case of environmental devastation in Eastern Europe before the political transformation) (Beck 2000: 219–220).

It would appear, also from other assessments of the state of global society (cited below), that we have entered upon an era of identifying and politically managing risk. We must confront the products of modernity – global networks of interdependence linking humans to the non-human world, involving a potential for destabilisation. Politics, societies, legal systems, belief systems and ethics should be better adapted to handle the dynamics of changes introduced by laboratories.

The author of *Risk Society* draws attention to the necessity of discussing the main assumptions of modernism anew. To take the new dimension of risk into account, we must first of all redefine the concept of nature, after considering the way in which nature is currently becoming man-made in its interaction with industrial systems. The notion of knowledge as such also requires redefinition. As Beck writes, we require a new theory of knowledge and a new understanding of the empirical (Beck 2002: 275–276). Currently, knowledge is largely becoming an element of politics. As it transpires ever more often, that which is deemed empirical and unequivocal can be used as a tool in the game of domination. In an era of risk, the environment must also be a domain of sociologists, not only of experts from science. The *de facto* social process of defining public problems should be made explicit. The question of the unintended consequences of the success of technoscience must not only be raised and discussed; it must also be stressed that risk (being virtual and constructed) is often subject to political instrumentalisation. Doubt and uncertainty are also produced professionally and on demand. Experts (whom we obviously cannot do without) are increasingly more often perceived as engrossed in conflicts of interest and lacking independence.

Beck also redefines the notion of the collective. In his book *Macht und Gegenmacht im globalen Zeitalter: Neue weltpolitische Ökonomie* he appeals for a global cosmopolitanism which would serve as an alternative to the political and economic system in place. In his opinion, whether we want to or not, we are witnessing the emergence of a new level of the critical meta-game for power – the transnational level. As before, it is being shaped by the most powerful agents of globalisation, or else, in certain domains, it is not normalised at all. The current policies of many states, concentrating on increased production and market competition, only intensify this risk.

In Beck's opinion, when constructing a new global order, we should leave state structures intact. In fact, the role of states should be strengthened. It is only through alliances and cooperation that individual countries will have the opportunity to take an active part in the meta-game for power. Instead of making errors individually, or competing with each other while subject to manipulation and lobbying by industrial actors, states should cooperate. Federations, such as the still evolving European Union, can provide a larger market, more extensive cost amortisation and an increase in sovereignty vis-à-vis global policy actors.

Apart from strengthening state structures, one should strive to develop basic, democratic institutions and procedures at the transnational level. As Beck states, we must fill the institutional void of the transnational domain, creating global laws, market and trade regulations as well as effective courts of arbitration. Furthermore, we also require global media as an arena where we are free to articulate the problems of public policy, creating global, informed citizens, who will decide which innovations shall be permitted in the future.

Beck's cosmopolitanism is an ambitious project, postulating widespread and thorough changes. It is also extremely utopian. In one commentary, we read that his cosmopolitan politics is "an idealist politics, without widespread support or any chances of coming to power" (Dybel, Wróbel 2008: 80). However, other prominent sociologists also reason in a similar utopian vein. For instance, Zygmunt Bauman notes that "The so-called 'globalisation' of the planet has been a purely negative process to date. 'Globalising' on an enormous scale at a frantic pace are finances and investment capital, the commodities trade, information networks, drug traffic, mafias, terrorism… Meanwhile positive globalization has not yet begun in earnest, namely the globalization of civil control, legislation, the judiciary, and above all of ethical norms which should give purpose and direction to all the rest" (Bauman 2007: 362). Bauman underlines the necessity of creating a global political and normative arena in the positive sense. We need transparent global regulations and transnational institutions, also in the context of monitoring technoscience.

In Beck's view, the construction of a cosmopolitan order does not need to be undertaken in the name of the rationalist ideals of the Enlightenment, nor in that of an utopian universal solidarity. As he very cogently states, it would be extremely hard to construct global ties while appealing to an abstract concept of solidarity. A cosmopolitan world order can be created as part of an attempt to obtain concrete gains in the meta-game for power (Beck 2005: 295). One of the basic gains for all of humankind would be a real, transnational reaction to the problem of risk (mainly environmental risk). In his project, Beck redefines the collective as "a community of fear", of solidarity in the face of global threats, generated by ourselves. This awareness of a new type of threat can become the foundation on which a planetary community of fate can be built (Beck 2005: 316). The author of *Risk Society* claims that we have already been confronted with similar situations in the past, when only events such as the Holocaust, Hiroshima and Chernobyl motivated humankind to take effective action transcending local interests. It is precisely in the face of global catastrophes that we created the first institutions aiming to protect humankind and the planet as a whole. One of the most important challenges we face in connection with risk is therefore to include climate policies in "great" politics (Beck 2009: 100). This could prove extremely fruitful, bringing the world together and forcing the creation of an effective network of transnational institutions. (As we shall see, in this respect Beck and Latour are of one mind, and moderately optimistic.)[13]

What is interesting is that other researchers also share Beck's hopes in this regard. For instance, the German philosopher Peter Sloterdijk claims that global warming can force societies to enact radical political changes, and even achieve economic growth (cf. Kutyła et al. 2009: 103; Sutowski 2009: 276). Similarly, George Soros, an economist and billionaire, sees hope in the fact that problems related to climate change could lead to the deconstruction of the erroneous assumptions of market fundamentalism and become a force of change within the world economy. Counteracting global warming shall require such immense investments that it shall replace the current trend in western societies to consume without producing.[14]

More and more analysts stress that "green", "low-emission" policies, introduced in response to environmental threats, could become an opportunity for the economy and labour markets. One of those analysts is Nicholas Stern, former chief economist at the World Bank. In a widely discussed report commissioned by the British government, Stern claims that capitalism could actually flourish in an era of ecological restrictions. "Green" capitalism would require new technologies and so create many new jobs (Stern 2009; cf. Beck 2009: 115).

Stern's argumentation is developed against the backdrop of traditional modernist and Enlightenment values. He does not question the need for humankind's further economic growth, consumption and investment. In that respect, Stern's position is strikingly at odds with the other positions we have discussed. The report in question appeals to the values of the capitalist era, demonstrating how ignoring climate change could endanger the world's economic growth.

---

13  Unfortunately, there is no place for a similar discussion within Polish politics, though the notion of sustainable development can be found in the Polish constitution of 1997 (article 5). As Adam Ostolski writes: "Poland and other 'new' European countries are conducting a cowardly policy with regard to climate change; a vast majority of the media are engaged in a nonchalant 'denialism'"(Ostolski 2009: 402).

14  See www.pbs.org/moyers/journal/10102008/watch/html, accessed 30.07.2009.

Stern expounds the advantages of taking action to stabilise the climate in the language of economic risk, while also estimating the cost of such actions. Among other things, he examines the emergent $CO^2$ emissions market as a favourable stimulus for the global economy.[15]

In Beck's view, our understanding of politics must also change in reaction to the problem of risk. When following modern scientific controversies and public debates concerning innovation, we can see that science has become a "self-service store for financially powerful clients in need of arguments" (Beck 2002: 267). The high level of commercialisation and the ubiquity of conflicts of interest inhibit the independence of expert opinions. Due to the unavailability of unequivocal expert opinions, not only ordinary citizens, but, above all, politicians, are often forced to make snap decisions in a setting of uncertainty and fragmentary knowledge. Political reflection in the era of the risk society should confront this fact, taking it into account in modern models of management, governance and influence. These models should therefore take into account the inexpungible uncertainty that accompanies political decisions, and consider the potential ambivalence of expert opinion.

In Beck's view, the actual making of important decisions has also shifted considerably. These are no longer made as part of a public, democratic political debate, but in the sphere of what may be called "sub-politics". As he writes: "The forms of political involvement, protest and retreat blur together in an ambivalence that defies the old categories of political clarity" (Beck, Giddens, Lash 2009: 37). Sub-politics means shaping society "from the bottom up", through channels and mechanisms different from those hitherto relied on (cf. Dybel, Wróbel 2008: 65–71). According to Beck, the political domain has shifted into that of corporations, where the binding decisions are made. Industry turns out to be a crucial area of sub-politics. With regard to technoscience, the real entities shaping its future are currently various types of consortia (pharmaceutical, biotechnlogical or other), privatised research institutions, defence lobbies and also, though to a decreasing extent, the most powerful superpowers.

---

15 Let us explain: an emissions market depends on defining limits for carbon dioxide emissions, then providing individual actors with permissions to emit. Unused emission limits can then be resold. Currently, the largest functioning system of this type is the European Trading Scheme (ETS). Let us note however, that emission trading is merely an outline of a market that would require extensive regulation and reforms. For instance, it struggles with the issue of apparent reduction of emission levels, since it is difficult to estimate the level of reduction for planned emissions, or verify it.

Let us note that traditional democratic politics are characterised by a long path towards the implementation of decisions, which leaves ample time for any eventual corrections, discussions or debates. However, as Beck demonstrates, within the realm of sub-politics (e.g. within scientific research or medical practice) we are faced with a specific immediacy. Implementation is instantaneous; there is no need for the prior legitimisation of a decision, while the effects will be dealt with after the fact. Within a risk society, laboratories and research institutes are political centres.

Therefore, if we are to create institutions that monitor the future of the collective and enforce norms of responsibility on a global level, they would have to be placed at the sub-political level, where the crux of the problem seems to be. Humankind has not yet developed institutions to rein in the dynamics discussed here (besides the omnipresent requirements of market rationality). It seems we have to agree with Latour, who writes metaphorically that, for all intents, the collective has not yet been collected. It is an arena of chaotic, undirected transformational processes. We therefore need institutions or procedures that make it possible to control sub-politics.

On the one hand, sub-politics is conducted beyond democratic control in non-transparent fields (industry, corporations). On the other hand, a cosmopolitan society allows individual consumers to have real political influence. According to Beck, ordinary citizens using particular technologies or gadgets are also actors of sub-politics. In a society of global interconnections, the domain of private decisions therefore also becomes political, for instance through a refusal to buy certain products.

As Beck claims, civic movements are therefore also becoming increasingly important players in the global game for power. However, they are definitely not aided by excessive specialisation. The goals pursued by these movements are so fragmented that they cannot develop a common platform and understanding (Beck 2005: 152; cf. Passmore 2002: 625).

Among civic movements, we can observe diverse groupings: ecological, feminist, human rights and workers' rights, consumer groups, alter-globalist as well as various non-governmental organisations. Each of these movements has recourse to its own methods and often also effective means of exerting pressure, such as boycotting certain services, alliances between human rights groups, organised strikes as well as the already mentioned refusal to purchase. What is important, the legitimising capital of civic movements or NGOs decidedly exceeds that of other actors on the global stage, such as corporations, banks or even superpowers. The role of transnational and civic movements is systematically increasing. Some of them are even starting to achieve significant successes. This is underlined by Beck, who, referring to Greenpeace, notes that it is successfully "staging a worldwide, mass civic protest using instruments from the media age" (Beck 2009: 110). Civic movements should therefore become active participants in an open discussion on innovation, the direction of development of technoscience, and by the same token, the future of the risk society.

To sum up, let us try to briefly present Beck's most important theses and postulates:

1) the problem of modern risk requires public awareness and urgent political solutions; unfortunately, science remains systemically blind to these issues, unable to perceive its own role,
2) in reaction to the problem of risk, the main tenets of modernism should be redefined: nature (as our construct), knowledge (as subjected to political instrumentalisation), science (as lacking a monopoly on rationality and provoking unwanted side effects), the collective (cosmopolitan, created in reaction to the risk of global destabilisation, e.g. climate change), politics (conducted in conditions of uncertainty, faced with the changing status of professional expert knowledge),
3) we must devise a democratic means of controlling the sphere of sub-politics, where the decisions shaping the future of the collective are taken. We need transnational democratic institutions allowing us to suitably direct subsequent transformations of the risk society: law, effective market regulations, global media,
4) ordinary citizens and civic movements should be included in the debate on risk.

## Latour's "politics of nature"

The political ecology of Latour, co-founder of the actor-network theory, is a direct reaction to the problem of the unintended consequences of innovation and the risk generated by the interventions of technoscience. Its detailed programme is presented in *Politics of Nature* (Latour 2004). As Nicholas J. Rowland writes, this is one of the works which demonstrate that the results of science and technology studies can also prove useful outside of their domain (Rowland 2005: 953). According to Latour, one of the main concepts of the actor-network theory, namely the "network", a tangle of surprising interdependencies between heterogeneous elements, is an ideal equivalent of Beck's "risk" (Latour 2003: 36). The increase of risk and lack of control result from the length and complexity of global interdependency networks.

In Latour's view, the unpredictable consequences of the interventions of technoscience and industry, in particular those which prove catastrophic, can no longer be viewed in isolation from the hybrid creations introduced through the market into our collectives. The role of researchers, engineers, entrepreneurs and technologists can no longer be neglected, since they not only incessantly create new networks, but also actively participate in maintaining their existence. Laboratories are therefore a domain in which politics play themselves out.

As the author of *Pandora's Hope* writes: "Science, technology, markets, etc. have *amplified*, for at least the last two centuries, not only the *scale* at which humans and nonhumans are connecting with one another in larger and larger assemblies, but also the *intimacy* with which such connections are made" (Latour 2009a: 61). Each day, we learn from the media about both direct and far-reaching connections between heterogeneous elements: science, morality, religion, law, technology, finance and politics. We build networks on an ever greater scale, our interventions reach deeper and deeper (it is one thing to use simple tools, another entirely to genetically modify organisms). We have in no sense emancipated ourselves from nature, but non-humans are ever more deeply involved in our practices. In that sense, modern man is intensely entwined with nature (which goes against the opinion of philosophers as to the separation of civilised man from nature). Ecology therefore cannot be based on hypocrisy – postulating the autonomy of nature separated from man.

According to the co-founder of the actor-network theory, current pro-ecological policies, enacted in the interest of non-human actors or the rights of future generations, remain quite ineffective.[16] As we read in the *Politics of Nature*, the actual political practices of ecological movements stand in stark opposition to the theories or self-presentations of ecologists (such as deep ecology or the aestheticisation of nature). Latour discusses the following differences (Latour 2004: 20–22):

1) ecologists think they are speaking about nature, while the actual objects of their discourse are the hybrid links between various elements, relationships

---

16 Obviously, this is above all ineffectiveness as compared to the enormity of the actions still waiting to be implemented. Non-governmental ecological organisations have many successes to their name. For instance, as a consequence of Greenpeace's actions, France desisted from nuclear testing on the Mururoa atoll, Australia banned whaling, while some European countries banned lead additives in petrol. In 1995, Greenpeace also successfully pressured Shell into abandoning its plans for a drilling platform at the bottom of the Atlantic ocean.

and networks in the process of being constituted, linking together instruments, regulations, consumers, customs, animals etc.;
2) though ecologists claim they are protecting nature from humankind, they are in reality intensely weaving humans, and the human, into diverse contexts. They do not seek to protect nature, but rather to raise the issue of necessary control over an increasing diversity of entities of a novel type and their future;
3) political ecology claims to defend nature for nature's sake, freeing itself from human egotism and imperialism, while it in fact makes its case mainly by referring to the good of humankind, future generations and aesthetic pleasures. It is *de facto* about the conscience of a small number of specifically chosen people: educated and well-off westerners. Moreover, political ecology cannot protect nature for nature's sake, since in reality it is absolutely incapable of defining a *common good* for an abstracted "nature";
4) political ecology states itself to operate within a System defined by the Laws of Science; at the same time, it is constantly involved in numerous scientific controversies in which experts are often unable to reach unequivocal consensus. In fact, and rather fortuitously, ecologists do not really know what constitutes this System and are unable to unambiguously define its links, consequences and causal relationships. Rather, their role serves precisely to underline this lack of certainty and knowledge;
5) when searching for scientific models describing the hierarchical order of nature, ecologists are often surprised to discover links and relationships that are chaotic, sometimes local and sometimes global, impossible to fit into hierarchies;
6) ecologists claim that their discourse concerns the Whole, while they are in fact always concerned with specific biotopes, situations and events. Moreover, they are unable to present, in an integrated and hierarchical manner, what it is they are concerned with. And again, this is precisely the greatest virtue of political ecology: that in the end, it does not present any absolute order. That within its domain, the smallest bit, such as baby elephants or a few whales trapped in ice, can prove to be extremely significant;
7) as ecologists themselves believe, they represent the most important political challenges of the future and are gaining in political strength. Meanwhile, ecological movements are marginalised all across the board – and it is so because of their own theoretical assumptions.

Latour argues that the way ecologists present their political goals, anchored in the traditional (essentialist) manner of thinking about nature, politics, science and technology, is in many ways erroneous. Let us note that thinking about the need, as such, of monitoring technoscience is often based on the conviction that the introduction of appropriate interdictions, top-down regulations and institutions guaranteeing the effectiveness of any eventual sanctions shall in itself constitute a solution. An interesting example of one of the first controversies concerning technical innovations, as well as attempts to control technology through top-down decisions, was the opposition to the interdiction of crossbows. Such a ban was introduced by the Second Council of the Lateran in 1139 (cf. Stankiewicz 2010: 197). This may have been one of the first (unsuccessful) attempts to control innovation through a simple ban. As it appears, top-down, one-time interventions made after the fact are not able to suppress the dynamics of innovation. An invention does not boil down to a single object that can be destroyed or outlawed. It is always located within a wider network of dependencies. Innovation should not be identified with innocent, isolated gadgets that are either effective or not. Rather, technologies are extensive cultural, material, social and political transformations.

Democratic control of technoscience would need to be global, systemically planned and institutionally supported. As Latour notes, the first thing to transform would be the current nature of the public debate on innovation: it should be conducted using concepts that reflect the heterogeneity of the phenomena under discussion. Since innovations and non-human actors enter into our daily lives in such large numbers, they should become citizens and be politically represented (Latour 1991: 18). Latour therefore postulates building a democracy extended to things, the creation of a Parliament of Things, in which it would be possible to represent the processes involved in the creation of hybrids and the role of non-humans co-forming the collective (Latour 1993: 12, 142–145).

We should note that the postulate of actually considering the political role of non-humans cannot be reduced to merely speaking about them, to their representation and the articulation of their interests within human political debates (for instance by ecologists). Technologies, artefacts and medical solutions are also vehicles of political solutions, tools of power and oppression, surprising members of the collective, part of the *polis*, acting and introducing difference. Viruses and ecosystems interact with us and redefine problems previously considered uniquely human.

Latour postulates the replacement of the notion of society by that of collective. The collective is a distinct notion, since it implies dynamic links between people and non-human actors, whose role, according to this French sociologist, should also be taken into account. Non-human actors have co-created and actively co-create the foundations of the world we live in. They include not only artefacts and advanced technological systems, but also those elements that we would traditionally describe as material or natural. Latour uses the term *collective* in the singular, which does not imply that there is only one such entity. Within the domain of the collective there are those who refer to themselves as "we" (Latour 2004: 210).

In Latour's view – and Beck held a similar opinion – we also require an open re-examination of the basic, inadequate assumptions underlying modernism: 1) the belief in the autonomy of science, 2) the belief in the substantial, key role of theoretical research, 3) the assumption that it is necessary to maintain the salutary freedom of scientific discoveries, as well as 4) the belief in the unequivocal, conclusive function of expert opinion. Even calling upon super expert opinion does not give us certainty, since the complexity and global nature of interdependencies currently exceeds our possibilities of a cognitive overview.

As the co-author of the actor-network theory underlines multiple times, we must also abandon the philosophical dichotomy of values vs. objective facts. The essentialist concept of facts is in many ways defective. For instance, facts developed by laboratories cannot be questioned, which makes ethics powerless, condemned to adapt to the proposals carved out by science. According to Latour, the dualism between ready-made facts and the values assigned by man is a normative and political construct, blurring the actual processes involved in the stabilisation of decisions and the gradual institutionalisation of the parameters of reality.

The hybridity of the entities created thanks to the establishment of links between human and non-human elements remains blurred due to the purification processes described in Latour's *We Have Never Been Modern*. What precisely does purification consist in? In general terms, it signifies a loss of caution. Modernity created the institution of the laboratory, in which effective processes of constructing facts and fabricating innovation take place on a large scale. According to Latour, that era simultaneously introduced an ideological ontological division: Nature (the domain of objective, unquestionable facts) and Cultures, Societies (the domain of negotiable norms and values) (Latour 1993: 41). Due to this, the construction of facts and artefacts through the creation of networks between diverse elements (material, social, political and symbolic) has ceased to be visible. It was decided that the role of technoscience does not require ethical or political thought, since laboratories create either "pure" natural facts or innocent technological gadgets. They are not hubs of social change or centres of power. However, in Latour's view: "In spite of its transcendence nature remains mobilizable, humanizable, socializable. Every day, laboratories, collections, centres of calculation and of profit, research bureaus and scientific institutions blend it with the multiple destinies of social groups" (Latour 1993: 37).

If we wish to discover the true role of laboratories today, when our own creations slip from our hands, we should question the beliefs associated with the ideology of purification. This does not mean that we should reject the facts and innovations produced by technoscience. We should neither attempt to block technoscience, nor forego the professionalism of researchers and engineers. We must learn to differentiate between the connections that protect us, and those that can kill us (Latour 2010a: 61). We should remember that facts and technologies have their own history like all other institutions. And that once established, they should not be abandoned or neglected. Like all products of our practice, they require continuous and thoughtful attention, since they can always become problematic in a controversy, or be transformed during use.

As Latour argues, modern politics should consist in gradual, communal experimentation. But this experimentation must be cautious. As he writes, democracy is such a fragile invention that it should be built most carefully of all (Latour 2002: 9). It should be the art of governing without ruling, aiming to create a community of people and non-human actors which takes the form of a harmonious order – the cosmos.

Due to the problem of the ambiguity of expert opinion, the modern conditions for conducting politics are significantly altered. "Those who wait for absolute certainty before acting are living in the wrong time", writes the author of *Politics of nature* (Latour 2004: 263, footnote 17). As we may remember, Beck also admitted that we are today forced to make decisions in a context of uncertainty (Beck 2000: 217). This particularly concerns politicians.

However, according to Latour, the collective can learn through experimentation, while attempting to find the most judicious political solutions. All the necessary institutions, tools and skills already exist (Latour 2004: 235), all that is needed is to modify their character, creating new types of interrelations between them. The valuable skills that we already possess are:

1) the qualifications of scientists, mobilising non-humans professionally thanks to their instruments and laboratories, introducing discoveries and innovations, speaking out in favour of things, presenting alternative possibilities, ending controversies by stabilising facts and defining the parameters for the collective;
2) the skills of politicians, who by articulating points of view, introduce the multiplicity of positions held by relevant entities, provide a voice to reliable

witnesses, handle negotiations, explain interests, consult opposition opinions and viewpoints, identify enemies who should be excluded, and create consensus;
3) the competences of economists, who provide a common language to heterogeneous domains, thus enabling hierarchisation, dispose of reasonable means of articulating preferences, since the rational, provisional economic balance sheet is the best available method of eliminating incommensurability, allowing to find optimal solutions;
4) the skills of ethicists, who above all arouse a feeling of uncertainty, forcing us to consider non-humans as ends and not solely as means. Ethicist reopen already closed debates in the name of the right of appeal. From the viewpoint of ethicists, the collective in its current shape is always a dangerous, provisional artefact. Scientists and politicians are too quick to exclude when defining the boundaries of the collective, while economists are quick to ignore what they are unable to take into consideration;
5) the qualifications of administrators and bureaucrats, who ensure the continuity of the collective's public life, creating and maintaining archives, documents, data and research results. All these resources make it possible to sometimes reverse decisions, renew controversies or reconsider the current state of the collective;
6) the abilities of diplomats, who conduct negotiations despite never being entirely sure in whose name they are acting and what precisely the collective that has appointed them is composed of.

In Latour's view, all the areas of competence listed above should interact with one another, and all the indicated groups should make decisions in common, at all levels. It is only the mutual "keeping in check" of scientists, politicians, ethicists and economists that prevents the world from devolving into chaos. Mutual criticism, as well as the pluralisation of positions within the dialogue on innovation, will make it possible to articulate many varied interests and to gradually reach a consensus. In the end, incommensurable values and goals need to be hierarchised if a political decision is to be made. We should, however, remember that no hierarchy is final.

As we can read in *We Have Never Been Modern*, the very disclosure of the fact that we are creating hybrids while later denying it, is sufficient to slow down these processes (Latour 1993: 10, 41, 141). The creation of stable associations in laboratories, and their simultaneous purification, should be considered jointly. When the creation of hybrids takes place officially, we will be able to subject it to democratisation. The facts or innovations emerging from time to time and aspiring to become part of the collective are always heterogeneous combinations of humans and non-humans: viruses and virologists, pharmaceutical company representatives and government agencies, pulsars and radio-astronomers or lions deeply associated with the Masai and their interests. We should not predetermine their ontological status while controversies are still raging, ask whether they are real or fictional, whether they are artefacts or facts of nature (Latour 2004: 171–172). Stabilisation will only be reached as an effect of wide-ranging processes, finally "The prion is indeed responsible for mad cow disease; the minister of health is indeed responsible for the deaths from blood transfusion; God is not to blame for the earthquake that destroyed Lisbon" (Latour 2004: 179).

The collective should react to all innovations with attention, sensitivity and caution. It is necessary to decide what tests and experiments the new entities should be subjected to, in order to understand and socialise them, making them elements of our world. It is also necessary to decide who shall represent them. Finally, we must gauge whether they will prove compatible with all that already co-forms our reality.

Quite obviously, according to Latour, we do not dispose of ready-made criteria to unequivocally define who should make all these decisions. One thing is certain however. When dealing with pulsars or viruses, the decisions should not be taken solely by astrophysicists or virologists, but by everyone who might have an interest in the consequences of the introduction of a given innovation. We can see here that the author of the *Politics of Nature* is clearly opposed to the inherent paternalism of experts.

But how is it possible to exhaustively list all those whose life will be deeply transformed if we introduce prions, pulsars or even avian flu? Unfortunately, the collective is always poorly informed in this respect. Appealing to the decisive voice of experts, speaking in the name of Nature, was merely a political shortcut. Negotiations aimed at reaching a consensus must take place between all the groups mentioned above: politicians, scientists, ethicists and economists. This implies that at the end of negotiations, we must once again redefine what we understand as "we". Added to this, according to Latour, an important condition in such a situation must be that compromise is reached publicly, explicitly, in a transparent, archived and documented manner. Thanks to this, the costs we bear, what we exclude, and the losses we subject the collective to, become apparent.

Making his political ecology project more precise and thereby also indicating the central areas of modern politics, Latour writes: "it is hardly likely that anyone can unify such disparate groups as those which affirm that the world

## B. Proposals

is made up of atoms and those which await salvation from a God who created the world six thousand years ago; those which prefer to shoot down migrating birds rather than belong to the European Union; those which want to develop gene therapy to cure their children, against the advice of biologists, if necessary; those which vote in Switzerland against the transformation of their rapeseed fields into a laboratory annex; those which oppose cultivating human embryos and those others, associations of victims of Parkinson's disease, which expect the same embryos to provide a cure. None of these members of the collective wants to have an "opinion" that is personal and disputable "about" an indisputable and universal nature. They all want to decide about the common world in which they live. Here ends the modernist parenthesis; here begins political ecology" (Latour 2004: 130). Summing up, Latour's politics of nature project postulates:

1) exposing the fact that laboratories are places where political decisions are made. The renouncement of purification practices aimed at slowing the chaotic multiplication of new proposals;
2) the articulation of the interests of non-humans within the political sphere, the creation of a Parliament of Things, taking into account the active political role of non-human actors,
3) rejection of the paternalism of experts, admission of multiple points of view (including the representation of non-humans) in debates concerning innovation,
4) transformation of modernistic concepts: nature (as woven into our practices, linked to the human), politics (as action within a context of uncertainty and unequivocal expert opinion), technology (as dynamic networks of associations that deeply modify the collective's parameters – and not simply inventions that can be banned). Rejection of the concept of society (in favour of the collective), rejection of the fact vs. value dualism (in favour of the concepts of network, factish, quasi-object etc.), abandoning false beliefs in the autonomy of science,
5) transforming the foundational assumptions underlying ecological policy (e.g. monitoring the growing number of entities in the collective instead of protecting nature),
6) using the already existing skills of scientists, politicians, ethicists, economists, civil servants, diplomats: mutual criticism between these groups, pluralisation of positions within an open, democratic dialogue on the future of the collective.

## Transformation of the boundary conditions of the public debate on innovation

Science and technology studies as well as the cognitive sciences dealing with the issue of situated and distributed knowledge demonstrate how objects, scientific instruments and technologies open up new fields of action and enable the externalisation of cognition and the obtainment of effective solutions. The actor-network theory emphasises the meaning of non-humans, which co-create social ties, moral norms, the success of technoscience, and generally, the current shape of global society. More and more often, scholars in the humanities underline that things and infrastructures can play a political and ethical role. From these voices we can deduce that the role of artefacts, technologies and the surprising consequences of heterogeneous associations should be taken into account within modern politics. How can we achieve this?

On the other hand, Beck's conception initiates a new stage in the discussion concerning risk. It modifies the fundamental boundaries of scientific debates on modern systemic risk, ecological issues and the premises guiding the actions of NGOs (Strydom 2002: 29). Unfortunately, it would seem that Beck's insights do not provide concrete political recipes. This is partly due to the nature of the problem itself – as Beck explains "Risks only suggest what should *not* be done, not what *should* be done" (Beck 2000: 218). However, the question of what should actually be done still requires an answer.

Ruth Levitas, examining the problem of risk, writes: "we need a sociological analysis which places at the centre of its interrogation the relationship between the accumulation of capital and the accumulation of danger, and which focuses on the real, as well as the discursive, production and distribution of risk" (Levitas 2000: 205). Instead of repressing social inequalities, as class societies did, the main task of the risk society turns out to be reducing and legitimising uncertainty. The key political question in the era of risk is whether a fairer distribution of technological threats is possible. However, who should decide this, and based on what criteria? Modern conflicts focus on the distribution of risk as well as the distribution of responsibility. In the meantime, whole sectors of the economy have arisen, basing their profits on identifying and eliminating threats. We have no other solution but to learn how to recognise the political, and above all the economic contexts in which technologies are developed (Levitas 2000: 207; Beck, Giddens, Lash 2009: 18, 75).

The political institutions of the modern world should be modified so as to be capable of facing increasingly more frequent controversies through constructive dialogue. We must find a formula for a "technological democracy" that questions the division between laypeople and experts and allows for the creation of hybrid forums: experts, ordinary citizens, politicians (Callon, Barthe, Lascoumes 2009). Hybrid forums should therefore provide conclusions for political action within a context of risk, in conditions as democratic as possible.

In the actor-network theory and the risk society theory the fundamental parameters of modernity are defined as contingent. If they are contingent, it means that they can be reorganised politically. The growing uncertainty of the future awaiting us should inspire us to act. The boundary conditions of the collective must be modified. We must however first identify the potential intervention points that could bring lasting changes to the current system (Levitas 2000: 209). One of the key, and most often mentioned elements requiring transformation is precisely the form of public debate concerning innovation. As we have seen, both Beck and Latour underline this fact. Let us therefore consider in detail the proposed changes regarding this debate, as well as their real chances of being implemented. We shall strive to answer the following questions: which debates exactly are we talking about? Who should take part in them? Is it possible to sow the seeds of these solutions? On what basis? Finally, what are the costs, difficulties, or unresolvable dilemmas that we are faced with in the implementation of these types of projects?

The rationalist political ideals characteristic of the modernist era (where politics is understood as engineering) do not even allow for the existence of unresolvable political problems. Meanwhile, the thinkers referenced in this text attempt to convince us that the assumptions behind technocratic administration fail utterly when it comes to the risk society and the uncertainty resulting from the complexity of global interconnections. As we shall see, the proposals discussed below for the transformation of the boundary conditions of the public debate concerning innovation represent an anti-technocratic position.

Technocracy, or government by technical experts, presupposes that scientific and technical competence can be substituted for political rule, while bringing about a future in line with the values articulated by experts. As Joanna Kurczewska underlines in her work *Technokraci i ich świat społeczny* (Technocrats and their social world), technocratic ideology presupposes that social relationships can be modelled on the relationship between man and nature (appearing in the domain of production). Moreover, the technocratic outlook is characterised by "the intellectual terror of the (eschatological) future". We are faced with it when the prognostic abilities of scientists and engineers, as well as their vision of the future, serve a legitimising purpose (Kurczewska 1997).

The risk society is therefore forced to reject the technocratic model of instrumental rationality, leading towards unequivocal conclusions. The monopolistic position of experts is questioned here. In its stead, the possibility of creating new conditions for public debate on the future of the collective is considered. Its participants are to be chosen based on criteria that the citizens themselves find valid. Despite the fact that experts do not agree among themselves in public debates and are unable to achieve a consensus, politicians still need efficient procedures for making quick decisions. As it would seem, these procedures should above all be democratic. Therefore, it is necessary to define the manner of debating, voting and approving solutions, as part of the procedures for introducing innovation (Beck, Giddens, Lash 2009: 48–49).

The idea of non-expert public participation in making decisions about science and technology has a history which starts at the end of the 19th century, as this subject did not really come up before then. In the 19th century, scientists fulfilled hybrid roles: intellectual, political and civic. They saw no need to involve laypeople in the process of giving direction to science and technology. The first bodies to enable this type of participation emerged in a non-academic domain, namely social insurance, as part of the occupational disease section of the International Labour Organization (Lengwiler 2008: 189–191). During the inter-war period, the participation of German and British scientists in developing the chemical weapons used during World War I met with a critical reaction from the public. This motivated states to create institutions to oversee science. At this time, both the interventionist United States and the USSR saw a strong politicisation of science and technology, with politicians taking part in the processes of monitoring these domains.

After World War II, until the 1970s, science went through a process of autonomisation, being solely subject to internal auto-regulation. Basic research was dominant, while claims for public participation were rare (Lengwiler 2008: 193). It was only at the end of the 1960s and in the following decades that the problem of the public role of technoscience was openly raised by ecological, feminist and pacifist movements (e.g. nuclear weapon opponents). Calls arose for the participation of ordinary citizens in the decision-making process related to scientific research and technological innovation. This was accompanied by a significant loss of public confidence in experts (in particular with regard to problems caused by the BSE epidemic as well as environmental and food crises).

Reflection regarding the democratisation of technoscience continued to evolve. Initially, questions were asked about whether new technologies and inventions threatened equality between citizens. Does technology not limit basic human rights? What values are materialised through innovations? The assumption that the intentions of the creators of innovation translate into the consequences of technology in a predictable manner were questioned. Since the 1980s, attention has continually been brought to the unavoidable uncertainty of the consequences of implemented changes (Nahuis, Lente 2008: 561–565). As a result, projects aiming to democratise technoscience have above all concentrated on how to ensure the possibility of democratic, consensus-building decision processes, in which responsibility is equally spread among the various groups impacted by the introduction of a technology. Currently, citizen participation concerns both the unintended consequences of inventions and innovations (above all related to nuclear energy or biotechnologies) and early decisions defining the appropriateness of research projects (chiefly within the biomedical sciences) (Lengwiler 2008: 188).

The attempt to widen and democratise public debate concerning technoscience is closely linked to a movement called "Public Understanding of Science" (PUS), born of science and technology studies. PUS first appeared in the 1980s in reaction to the growing number of controversies about technology and the increased distrust of the public towards experts. Let us take a closer look at it.

Since many of the controversies undermining confidence in experts erupted in Great Britain (cf. Yearley 2005: 113–114), in 1985, the Royal Society published a report entitled *The Public Understanding of Science*, which demonstrated the urgent need of promoting public acceptance of science in order to improve its relations with society.[17] Like public opinion, the British government had gradually lost interest in science in the 1960s, 1970s and 1980s, since the latter had limited potential as far as quick commercial rewards were concerned. This motivated scientific institutions to launch studies concerning the problem of trust in expert opinion. The birth of the research area discussed here is one of the indicators of change in the social role of science at the end of the 20th century. As Fuller writes: "If science has a public relations problem, it is not due to public hostility or even indifference to science. Rather, it would seem that science is being taken off its pedestal and shifted to some other place in our culture" (Fuller 2006: 2).

In the beginning, Public Understanding of Science mostly attempted to improve public perceptions of the domain. For instance, the first PUS studies showed that medicine, identified with public health, was considered the most scientific domain by laypeople. As it turned out, public opinion sees science through the prism of its knowledge about medicine (Durant et al. 1992). PUS research also showed that the common understanding of science is not so much

---

17 Alan Irwin and Brian Wynne discuss the incorrect assumptions made by the authors of this report (Irwin, Wynne 1996: 214–219). These included: a belief in the autonomy of science vis-à-vis society; the assertion that public opinion is a collection of atomised individuals; as well as the assumption that the scepticism of citizens derives from misunderstanding and ignorance.

about understanding theory, as about the valuation of scientific institutions that ordinary people come into contact with (Yearley 2005: 127; Durant 2009: 8). Attempts were made to use these findings to improve the image of experts. What is interesting is that PUS also devised means to regulate the controversies around technoscience, attempting to increase public acceptance for particular innovations and ways of legitimising new technologies. Often, this consisted in attempting to initiate open, public debates on the risks and benefits related to particular proposals for change.

In Irwin and Wynne's opinion, it is not only ordinary citizens who display ignorance towards scientific issues. Scientists also do not understand laypeople, or even their own socio-political role, by not displaying enough auto-reflexivity (Irwin, Wynne 1996: 10, 216). As these researchers convincingly argue, the practical skills and traditional domains in which laypeople function often allow them to make pertinent judgements on controversial issues. In fact, often more pertinent than expert opinion.[18] However, it is worth remembering that the image of laypeople themselves is often culturally conditioned and should not be idealised. The question of the distance separating laypeople from experts also should not be studied solely during controversies, when scepticism towards science is already present. Laypeople normally do not question expert decisions, in most (non-controversial) cases maintaining a high degree of trust in specialists.

As part of the reflection on risk and technoscience we can observe the gradual emergence of a defined attempt to guarantee the possibility of open debates on the role of innovation. In line with this, apart from experts from the natural sciences, government and industry representatives, sociologists, ethicists, advocates of the rights of future generations and non-human actors should also be included in the discussion of controversial findings and technology. There is even a reference to the ideals of direct democracy here (Nahuis, Lente 2008: 565). It

---

18 There are interesting differences concerning the perception of laypeople within the PUS movement (Durant 2009). For instance, Wynne notes that the participation of laypeople should not be examined solely on the cognitive level, but also on the cultural and hermeneutic one. He considers laypeople to be more reflective than experts. On the other hand, Harry Collins emphasises the importance of routine, non-reflective judgements and reactions, both among experts and laypeople. Giddens and Beck neglect the dimension of cultural meanings – when analysing experts and laypeople they concentrate on the intellectual and rational layers. Interestingly, since the necessity of bringing the viewpoints of experts and laypeople closer together is noted ever more frequently, the notion of Public Understanding of Science has been replaced with that of Public Engagement with Science (PES) (Durant 2009: 18; Marres 2007: 761).

is assumed that credible advocates of various groups, or even citizens directly impacted by the changes introduced, are also to be given a voice (patients, users, activists) (Papadopoulos 2010: 183). The goal is not only to have experts justify and legitimise their opinions in front of a larger audience, but to truly expand the decision-making bodies.

We should note that in modern political philosophy there is a substantial debate between proponents of participatory and deliberative democracy.[19] At first glance, both models seem concurrent, yet there are significant differences between them. Let us examine these in detail.

Deliberative democracy theories assign a key role to public debate, where arguments are presented in a cogent manner, respecting the subjectivity of the participants, who are seen to be engaged in rational discussion. We therefore witness a clear attachment to the ideals of rationality. The criteria of rationality are provided by the normative contexts of the communities in which the deliberations take place. Debate should not be conducted for the sake of argumentation alone; aside from its consultative and informational functions, it should also serve as a basis for actual political decisions. Pluralisation and greater openness of dialogue should serve to increase the transparency of the arguments on each side, which should in turn facilitate the rational solving of a given problem. Proponents of deliberative democracy usually believe that it must be supplemented by other forms of democracy, for example procedural democracy (Król 2008: 171–172).

Deliberative theorists criticise participatory democracy models as incomplete. Low quality mass participation does not in itself guarantee adequate conditions for democracy, since it can lead to the emergence of arbitrary forms of political power (Held 2010: 306–307). The issue is therefore not only increased participation, but also raising the quality of participation. Care should be taken to ensure that impartial, equal participants of debates can openly state their arguments. Consensus should be achieved thoughtfully, and a critical outlook maintained. Discussion helps to uncover the unvoiced aspirations guiding our actions (for instance, revealing whose vision of the common good stands behind political decisions).

Within technoscience studies, despite accepting many of the premises of the deliberative approach, there is often more emphasis on the simple expansion of participation than on the ideals of rational, idealised deliberation between

---

19 As David Held notes, the term "deliberative democracy" was introduced by Joseph Bessette in 1980 (Held 2010: 300). In the 1990s, we saw the term used within democratic theory (Lövbrand, Pielke, Beck 2011). One of the most significant pioneers of deliberative democracy theory is James Fishkin (Fishkin 1991).

impartial and cogent participants.[20] There are several reasons for this. Let us note that the deliberative concept of rational, public argumentation is incredibly utopian. It posits the possibility of an unequivocal settlement of burning issues through the use of transparent arguments, and presupposes a set of universalised criteria inherent to western rationality. In the deliberative model, expressing preferences without the ability of publicly and rationally justifying them is not desirable. Positions that are not supported by cogent, conclusive arguments are ignored. Furthermore, deliberative models idealise the participants of debates themselves. They are considered to be capable of impartiality, altruism and criticism. Finally, deliberative democracy theories approach the very problem of deliberation in a problematic manner – as the reaching of unquestionable, rational consensus.

Risk and technology scholars underline that instead of constructing rationalist, Euro-centric utopias we should rather reflect upon the nature of real debates taking place in non-ideal conditions. It is on this basis that we will be able to improve practices and transform institutions facilitating debate between actual citizens rather than impartial and altruistic experts. Debate participants often do not use rational justifications alone, but also examples, stories and gestures. What is particularly significant is that public debate does not necessarily lead to rational agreement, while still improving democracy. As I will argue below, the point of debate in the approaches analysed here is not final consensus, but rather a temporary, "robust fit" between the articulated values and viewpoints.

\*\*\*

As it follows from the postulates analysed here, public debate on the unintended consequences of laboratory discoveries should be designed as openly as possible. Let us therefore reflect on the voices it should include. On the one hand, a special role should be played by authorities from the realm of culture and intellectuals from the humanities: sociologists, philosophers, economists, cultural scholars, ethicists as well as religious authorities. These intellectuals should be able to adequately represent significant aspects of the functioning of technoscience as well as the condition of modern society, global and dynamically changing. Secondly, they should be able to rapidly identify problems relating to risk outside of the views formulated within the axiomatic assumptions

---

20 It is very doubtful whether Latour or even Beck admitted the existence of universal, supra-cultural norms for rationality. The sceptical relationship of STS and PUS towards rationalising deliberative models is linked to the overt questioning of the privileged status of expert knowledge.

of modernism. A sensitive reaction to the problem of risk requires the ability to question the logic of profit-making, to take into consideration the whole spectrum of alternative values and to question solutions implemented through market channels (often commonly seen as unproblematic).

Anticipating the unintended effects of innovations on social relationships and bonds as well as legal and normative structures requires an aptitude for "scenario thinking" (Beck, Giddens, Lash, 2009: 235). It therefore requires intellectuals with a "sociological imagination", as Charles Wright Mills famously termed it. Mills wrote that a sociological imagination is the "capacity to shift from one perspective to another – from the political to the psychological; from examination of a single family to comparative assessment of the national budgets of the world; from the theological school to the military establishment; from considerations of an oil industry to studies of contemporary poetry. It is the capacity to range from the most impersonal and remote transformations to the most intimate features of the human self – and to see the relations between the two" (Mills 2008: 54–55). Mills believed that this type of imagination consisted in the ability to understand the relationship between history and individual biographies within a given society. In relation to the problem of risk, the sociological imagination would need to be able to identify the surprising role of non-humans and to track the relationship between the possible effects of a given innovation across various ontological domains. Due to the scale and reach of the interventions already made by technoscience (in particular the biomedical sciences which introduce hybrid objects into the collective and manipulate what was previously termed "natural"), we also require thinking that is positioned outside of essentialist assumptions, and which is not anthropocentric.

As I have already mentioned, economists also seem to be irreplaceable participants in the public debate on the future of the collective. Perhaps we should even consider performing risk analysis first and foremost within the context of economic profits and losses. Such a move would allow us to conclusively close debates on values, which very often turn out to be irresolvable. As we have seen, Latour is partially an advocate of this type of solution, as is Cass Sunstein, an American legal scholar (Sunstein 2002, 2005). Sunstein underlines the fact that we should consider the possibility that we will often be unable to reach consensus between disparate views of life or visions of the common good. In such cases, reducing the debate to quantifiable issues, even if only provisionally calculated, can prove immensely useful.

Let us note however, that the idea of reducing debates on risk within the political context of the future of the collective to a purely economic calculation gives rise to many difficulties. For one, how do we gauge the security of future generations, the aesthetic dimension of the environment or the irreversible loss of species? According to many, they should have an infinite value. Comparing them, or for that matter placing an unequivocal value on them, seems impossible. However, despite these difficulties, the instrumental use of economic estimates in public debates on risk seems inevitable. Nevertheless it is important that priorities other than economic also be seriously considered in such discussions.

Finally, debates on technological innovations and risk should include lay voices. Let us examine this postulate in more detail. This closer examination requires defining the differences between two opposite models describing public reactions to the problems generated by discoveries and innovations. One is the paternalistic expert model, of which Sunstein is a proponent. The other model is based on debate and the engagement of truly interested parties. Defenders of the participatory model include Sheila Jasanoff and Brian Wynne, among others. Jasanoff and Wynne are representatives of Public Understanding of Science (Wynne 1992, 1995; Jasanoff 1995).[21] As it happens, the question of the unintended consequences of the success of technoscience, which requires political solutions, directs the thinking of researchers to the fundamental problems of politics.

According to Sunstein, we cannot rely on the opinions of non-experts in debates on innovation. In his view, laypeople are subject to many preconceptions when it comes to the issue of risk, for example neglecting the reciprocal ties between its different dimensions. Moreover, they evaluate risk solely on the basis of the available case known to them, while also relying on second-hand information without verifying it. Key political decisions concerning technological innovations should not be left to groups that do not possess professional knowledge, even when the effects of those innovations concern them. We can consider Sunstein's position as an example of the paternalistic technocratic approach.

But as it turns out experts are subject to the same biases regarding risk assessment as laypeople (Kusch 2007). Research done within science and technology studies highlights the fact that scientists also make incorrect assumptions about public opinion and lay beliefs, as well as science itself or their own roles. A lack of criticism towards experts can also prove much more dangerous than the biases of laypeople. Placing faith in the methodologies of their respective disciplines, scientists reify risk and describe it in the language of the natural sciences, while

---

21  It is worth noting that Wynne acted as an adviser on risk to the British government, the European Union, the UN, Greenpeace and the OECD, while Jasanoff works at the Harvard Kennedy School of Government, forming the elite of the US public administration.

neglecting its other aspects: legal, ethical, social, cultural or even metaphysical. Specialists also ignore the fact that the image of risk is always a certain construct, a model that emerges when complexity is reduced at the cost of neglecting certain interconnections. Experts see public opinion as a collection of atomised individuals (failing to notice the social mechanisms shaping common knowledge). Scientists do not perceive the socio-cultural determinants of world-view differences. They suppose that the ignorance of non-professionals results from a certain lack of intellectual capacity, while in fact these people have the same potential to use scientific knowledge. Experts also believe that laypeople are not really capable of discussing the form of society.

Meanwhile, there are more and more examples of the beneficial influence of laypeople in the monitoring of science (including influence on research processes as such). One of these has to do with biomedical research protocols for ways to cure AIDS. Activists engaged in behalf of AIDS patients have become actual, active participants in the process of amassing scientific knowledge concerning this disease as well as the therapeutic techniques and medications used to treat it (Epstein 1995: 409–410). Activists have been able to gain the status of credible participants in scientific debates, and have even been able to modify the understanding of what constitutes a credible conclusion within the domain of what we are examining here, namely research. They have also modified the nature of the patient-doctor relationship. As Steven Epstein notes, this example illustrates how difficult it is to claim that science is an autonomous discipline. Despite this being a specific example, the author notes that we can often observe similar phenomena in the biomedical sciences (Epstein 1995: 427). In the last decades of the 20th century, there was even a larger "expert-patient" movement, relating to other diseases. Today, we can say that these activists had a positive influence on medical practices and treatment procedures (Kerr, Cunningham-Burley, Tutton 2007: 386).

The expert model, paternalistic in its basic premises, does not allow for the inclusion of lay voices in the process of judging research activities, and presupposes, wrongly as it turns out, a high degree of autonomy for science. It assumes that social tensions related to the introduction of innovations can only, or mostly, be eased through various techniques of "enlightenment" of laypeople (Stankiewicz 2007: 93). This type of thinking is also termed the "positivist" or "deficit" model (since it presupposes a deficit of knowledge in laypeople, whom experts should simply try to educate) (Durant 2010: 205). Meanwhile, a factual, democratic alternative to the enlightenment model would consist in empowering citizens to co-determine the regulation of technological disputes. What form should this take?

## Concrete ways of expanding participation (and their respective weaknesses)

Frank N. Laird, an American researcher on technology management strategies, has attempted to find an answer to the question asked above. In one of his articles, he discusses the possibilities for expanding public participation in the democratic processes of implementing innovations (Laird 1993). Laird is concerned with tactics that can be implemented outside of government and administration. In his article, he compares two theories of democracy: pluralism and direct participation. Within both of these perspectives, real democratic processes are something more than just citizens voting and making decisions by proxy of elites. While pluralism cedes decision-making to groups, direct participation theory involves individuals. In both cases, what is significant is not just the chance to participate, but above all the quality of the participation, allowing citizens to thoroughly understand a given issue and guaranteeing an efficient use of influence (Laird 1993: 348).

Pluralism (a liberal theory of interest groups) sees democracy as based on the actions of and competition between organised interest groups which interact among themselves and with the government (Laird 1993: 343 and following). In our case, these represent the interests coalescing around a given technological innovation or ways of reacting to the consequences of laboratory interventions. These groups should be allowed to form freely and to obtain the information necessary within a particular context. They should also dispose of effective means of applying pressure, to influence the decisions of the government and administration. In the pluralist model, interest groups represent the interests of the citizens comprising them in a professional manner, while also aiming to attain their ends by hiring lawyers, experts etc.

On the other hand, direct participation theory assumes the full participation of individuals, their real participation in the decision-making process. It allows to take into account the educational effects of the individual strategies used in the participative process, never assuming that the interests of citizens are immanent and unchanging. I believe that this constitutes its main advantage over pluralism. The experience of participation transforms the positions of those involved, building up their engagement and understanding of community issues. If we were to facilitate true participation in the decision-making process concerning innovations, a side effect would be the development of responsible citizenship, which contributes to true democracy (Laird 1993: 346). Moreover, direct participation in the decision-making process also often guarantees that citizens perceive the results of a particular debate as legitimate. The direct participation of non-specialists in the decision-making processes relating to innovation therefore has important advantages. It not only democratises supervision over the development of technoscience, but also has a beneficial effect on democracy itself.

As an example of the application of pluralist principles, Laird discusses negotiated rule-making, which consists in working out a group consensus on a proposed administrative regulation by negotiating committees (Laird 1993: 350). Ordinary citizens are not part of these committees; they are composed of civil servants and representatives of interest groups (in well-planned cases, they should come from all such groups). The committees make it possible to arrive at a deeper understanding of the issue in question. Normally, they are constrained by a time frame for arriving at a decision.

One of the many forms of direct participation are so-called citizen review panels, composed of ordinary citizens selected by various means (Laird 1993: 352). An example of this is the Cambridge Experimental Review Board, a committee which recommended directions for DNA research in Cambridge, Massachusetts. In Laird's opinion, if these types of committees are guaranteed decisional efficiency, they will fulfil the requirements of direct participation theory, engaging non-specialists and allowing them to obtain the requisite knowledge on the issue discussed.

As Laird stresses, the key issues in both cases are the tactics for educating citizens participating in the decision-making process. They should be able to perform a participatory analysis of the issues they are examining, which means not only familiarising themselves with the facts but also being capable of analysing the issue itself, comparing various interpretations and evaluating their legitimacy (Laird 1993: 353–354). Participatory analysis during the process of forming one's *own* opinion requires a skilful, distanced use of expert knowledge. In this context, it is important to keep in mind the limitations of expert knowledge, as well as the fact that divergent opinions often exist between the experts themselves.

As part of the context discussed here, it is also worth mentioning other ways of promoting participatory action on the part of citizens, such as Participatory Action Research. One of the most successful examples of the implementation of participatory solutions to the issue of the introduction of new technologies can be found in Denmark. In 1980, the Danish Board of Technology[22] started to organise so-called consensus conferences, a form of the previously mentioned

---

22 A similar institution functions in Great Britain: the Parliamentary Office of Science and Technology.

citizens' committees. As part of these regularly organised consensus conferences, ordinary citizens with no formal ties to any of the interested parties prepare reports containing non-binding recommendations for the Danish parliament. Experts, as well as interested parties, have the possibility of presenting relevant information to these committees. However, reports are compiled exclusively by citizens sitting on the committees and taking part in the deliberations. The Danish citizen committees described here, created on the premise of the participatory democracy model, can even be seen as institutions allowing to regulate experts (Turner 2001: 125). Despite the incommensurability of the values represented by scientific knowledge and lay world-views, they allow for a debate between professionals and laypeople on an equal footing.

Fuller also writes on the subject of consensus conferences, also known as citizen juries or councils (Fuller 2006: 157–168). Such councils initially appeared in the 1960s in Germany and the United States. The issues raised in them range from cloning, genetically-modified food, and gene therapy to the information society and nuclear waste storage. These conferences usually last between 4–5 days. They have also been introduced in Japan and Europe. Other solutions of a similar type are deliberative polling in the USA and stakeholder dialogue in Europe (Lövbrand, Pielke, Beck 2011: 477).

Other institutions worth noting here are the science shops created at universities by the Dutch government in the 1970s. They carry out scientific research commissioned by trade unions, ecological organisations, NGOs etc. (Laird 1993: 357). One example of such research was the analysis of soil samples conducted by students at Utrecht University. A network of these types of institutions has currently been implemented within the European Union, while they have also appeared in Canada. Science shops should carry out free research for citizens, in line with the ideals of public influence on science and technology in a civil society.[23] They allow laypeople to work out their own, balanced opinions on disputed issues. Science shops are institutions that permit public access to expert analyses, an alternative to the often costly analyses prepared for governments, corporations or various lobbies. This type of project ensures a pluralism of expert opinions, but does not question their value.

Despite the many interesting attempts to implement solutions allowing citizens to freely and openly participate in the decision-making process concerning technological innovations and risks, opponents of the participatory model draw attention to the fact that the most vital public questions cannot be

---

23 See http://www.scienceshops.org/.

decided by "plebiscites".[24] Concerns of this type remain in accord with the paternalistic acceptance of the elite status enjoyed by experts, both politically and epistemologically.

However, we must admit that the participatory model faces several fundamental problems. In a discussion, compromise often seems impossible to achieve, in particular when significant differences of opinion or values emerge. In addition, putting too many issues up for debate or expounding upon too many arguments decreases the speed and increases the burden of decisional processes. Debate therefore cannot be employed to legitimise each and every political decision. We would therefore require criteria to select innovations or issues that *unconditionally* require an open debate. Unfortunately, there are no ready-made algorithms to decide this.

Real public participation in the decision-making process concerning innovation must take place within a suitably prepared institutional framework. Debates should channel the voices of interested parties in a balanced and transparent manner (depending on the weight of the interests they represent). Evidently, not all citizens can take part in all decision-making processes. I believe that the most popular public consultations performed today, namely those over the internet or the telephone, are still not fully satisfactory. What is more, not every case of participation in public debates takes the desired form (since we deal with persuasion, manipulation etc.). But ensuring the participation of representatives of different points of view in the political process of shaping the road of technoscience is only the beginning of the difficult struggle for the equal representation of diverse interests in a space that is free of domination (Lengwiler 2008: 196).

Social consultations and scenario workshops should be held sufficiently early, effectively and conclusively. As it stands, one of the main challenges is to devise successful methods for closing debates (Lengwiler 2008: 194). Accusations often appear that the conferences and committees boil down to discussions in an agreeable atmosphere, which smooth over conflicts around innovation instead of being a true forum for the independent articulation of opinions by various stakeholders. The status of decisions taken by consensus conferences is also a highly relevant matter. If these conferences are to fulfil their stated roles, their decisions should be binding or local authorities should at least publicly explain why the opinions of these bodies were ignored.

---

24 Prof. Tomasz Twardowski, a Polish biotechnologist, expressed himself precisely in this manner on the issue of opinion polls relating to Polish perceptions of genetically modified food (Stankiewicz 2010a: 31).

Sadly, there is also no certainty that ordinary citizens want to take part in the decision-making processes concerning innovation. It is possible that they do not require this type of responsibility, that they do not want to make decisions or even take part in consultative processes.[25] It is therefore hardly surprising that the antagonistic, paternalistic model of lay enlightenment remains popular, not only among experts and decision makers but also among ordinary citizens (Sismondo 2010: 178). It is worth underlining that due to the assumption of *voluntary* participation in consultation, the positions articulated within debates are reduced to those of the sides already initially interested and engaged in the issues (or at least positively inclined – in those cases where participants are chosen by lot, they can always refuse).

It is interesting to note that social consultations are becoming a more frequent phenomenon in Poland as well. They are performed as part of so-called environmental impact evaluations and mostly concern procedures of issuing permits to build roads, dams and waste incinerators (Okrasiński, Wasilewska 2011: 58). Environmental impact evaluations are performed as part of the UNECE Convention on Access to Information, Public Participation in Decision-making and Access to Justice in Environmental Matters, usually known as the Aarhus Convention. Sadly, the practical organisation of consultations leaves a lot to be desired, since it does not usually guarantee the actual participation of citizens in the debate concerning particular investments. In their article *Udział na niby* (*Pretend Participation*), Krzysztof Okrasiński and Agnieszka Wasilewska highlight the most frequent pathologies plaguing consultation in Poland. These include: 1) lack of access to information on a given project, or short periods for responses set by public bodies, 2) transforming consultation into pseudo-debate, serving more as a "safety valve" rather than a true debate, 3) the creation of claquers – artificial citizen support groups set up by actors interested in the success of a given project 4) situations in which investors scare off citizens with the threat of financial and legal action for "blocking the investment" (Okrasiński, Wasilewska 2011: 59–62).

In all certainty, creating real public debates on innovation and risk, designed in a democratically open manner, is a true challenge for practitioners. However, no other option exists. The antagonistic, paternalistic approach, ceding decision-making to experts and elites, is quite rightly criticised and rejected by proponents of democracy. The oft-used and excessively costly information

---

25 This is precisely the conclusion reached by research carried out across Europe in 2007 (Lövbrand, Pielke, Beck 2011: 489).

## B. Proposals

and education campaigns on the subject of innovation (in line with the deficit model), have little to do with true public debate (Stankiewicz 2010a: 36). I believe that we should in fact question the very point of organising these types of campaigns. In fact, the paternalistic model of lay enlightenment simply removes technoscience from the democratic decision-making process (cf. Yearley 2005: 119; Stankiewicz 2010: 218–219). As a consequence, the future of the collective is decided elsewhere: within the domain of sub-politics, in the laboratories of individual corporations, or is in fact not monitored at all.

What is important is that the participatory model does not need to assume that only one version of the common good is valid. The end product of debates does not need to be an unquestioned consensus. One can also think of political decisions in non-essentialist terms, without conceiving a pre-set order of best solutions. As it would seem, this is precisely the approach most often adopted within STS and the sociology of risk. It bears highlighting that there are no top-down standards or unambiguous algorithms for a "super-method" of achieving political consensus (Lövbrand, Pielke, Beck, 2011: 488; Braun, Kropp 2010). Just the uncovering of tensions and points of discord democratises politics. The exposure of differences between the various points of view of debate participants turns out to be a lesson for decision-makers. Fully engaged interest groups and citizens affected by the repercussions of a given innovation can play a substantial part in political decision-making processes. They are deeply anchored within the context of particular relationships, which will be reconfigured after any potential changes. They have the resources to learn as they act, which gives them a distinctive cognitive position, which in turn allows them to define the complicated issues related to a given controversy (Marres 2007: 776).[26]

I would therefore like to clearly emphasise that all we can achieve by closing a debate is a temporary, robust fit within the space of the interests negotiated and the values articulated. Similarly to what happens in the case of the underdetermination of laboratory practices, where the goal of research is interactive stabilisation (and not a final representation or truth) (cf. Bińczyk 2013a), in the case of the underdetermination of political practices, we should strive towards a relative, robust fit. The point of discussion is not reaching a final perception

---

[26] As one of the first thinkers to do so, John Dewey claimed that only public participation allows to satisfactorily define complex social issues, which existing democratic institutions are unable to deal with. He identified the importance of the unintended consequences of industry which should be subject to public monitoring (Marres 2007: 765 and following).

of the common good, or a rational, unquestioned consensus. Those would be unattainable ends.

Summing up, let us present the most important findings concerning the conditions for the transformation of public debate and the decision-making processes related to innovation and risk. Let us note at the same time that these projects are viewed as the most concrete attempts to democratise technoscience within the STS approach (Nahuis, Lente 2008: 578). They postulate the following:

1) public debates on technological innovation and the direction of scientific research should be designed as open, in line with the ideals of participatory democracy,
2) they are meant to enable representatives of various interest groups concerned with the consequences of a given innovation to articulate their views (in a space free of domination), or even the participation of ordinary citizens in the decision-making process,
3) they should take into account a wide spectrum of opinions, not only those of experts from the domain of the natural sciences, but also from the humanities (sociology, economy, ethics, culture studies, law), spokespeople for non-human actors and the rights of future generations, as well as those directly impacted by the consequences of the changes introduced,
4) debates should: be selective and relatively inexpensive (they cannot concern all issues and involve all citizens), be designed as conclusive and effective (binding for decision-makers), enable laypeople to arrive at their own conclusions within participatory analyses;
5) the goal of debate is not to discover a final vision of the public good, but to define a robust, temporary fit within the space of articulated values, divergent interests and perspectives.

## The Precautionary Principle, Technology Assessment (TA) and select examples of other institutional solutions

Apart from these postulates for the democratisation of public debates on laboratory interventions and risk we also find interesting proposals for monitoring technoscience at its very core. What remains relevant is the attempt to design solutions that allow for a democratic selection of the direction of research *before* it has even started. In this part of the argumentation, we shall present these solutions and discuss a few of them in detail. Without doubt, the most interesting modern proposals include institutions serving to evaluate technology as well as the Precautionary Principle.

As it turns out, some postulates, as well as wider political projects concerning possible ways of monitoring technology can already be found in the history of philosophy. For instance Werner Sombart, in his 1935 text *Die Zähmung der Technik*, proposed to convene "cultural councils" which would autocratically control technology through cultural criteria (Sombart 2001). Sombart's approach is an example of an authoritarian bureaucratic proposal to monitor technological innovation in accordance with priorities that are external to the market. Cultural councils would be composed of experts from diverse domains.

Neil Postman, criticising the technologisation of society, also presents a simple programme of institutional changes. He proposes to introduce technological education in schools. Such classes would present the history of technology and illustrate ways in which innovation brings about economic and social changes (Postman 2001: 183). According to the author of *Technopoly*, the purpose of this type of education would be to develop people's capacity to ask simple questions about the goals of technological progress. It is always worth being aware whose problems a given technology solves, who attaches importance to them, who profits from a solution and who will pay for it (Postman 2001: 49–53).

Although Postman's reasoning seems sound, an "educational" programme of reacting to the problems generated by technoscience must be deemed insufficient. Such a solution presupposes an effective emancipation of society through the spread of knowledge. However, the transformation of individual viewpoints and world-views through education is but a partial, and extremely slow solution. We also cannot be sure how effective it is.[27]

Likewise, Sheldon Krimsky, approaching the question of the contemporary condition of the biomedical sciences in the United States, articulates a certain number of political postulates in his work. In doing so, he concentrates on the problem of preventing the unintended consequences of the commercialisation of science, a phenomenon he examines. In the opinion of the author of *Science in the Private Interests. Has the Lure of Profits Corrupted Biomedical Research?*, a policy consisting in transparency and an increased exposition of the links between science and its sponsors alone is insufficient (Krimsky 2006: 318). Krimsky is a proponent of a much more decisive approach, aiming to eliminate conflicts of interest from academia. Scientists who hold shares in a company should be barred

---

27 Let us however note that the idea of introducing technological education at the academic level was implemented in the USA, starting in the 1970s. Science, Technology and Society (STS) programmes answered this very need. Currently, academic research programmes in science and technology studies often fulfil this educational function, understood as Postman intended.

from participating in research projects sponsored by that company. Furthermore, federal authorities should reward scientists who refrain from commercial activity. Finally, in the face of the encroaching commercialisation of science, Krimsky underlines the significance of tenure. It is hard to contest that tenure guarantees scientific independence and is conducive to academic freedom, facilitating open criticism. In Krimsky's view, this institutional solution brings about positive epistemological effects. The abolition of tenure would be an undesirable step down the path of the dismantlement of the traditional ethos of academic science.

I would like to note that the preservation of the autonomy of science in the face of market requirements does not yet constitute a solution of the problem of monitoring our future. On the contrary, it would seem that, as already mentioned, a serious inclusion of a wide array of values (including non-expert ones) in debates on the desired development of technoscience would require limiting the paternalism of experts, and by the same token, questioning the autonomy of scientific institutions to some extent.

Within the context of the need to monitor the development of technoscience, interesting postulates on the need to transform the mechanisms of self-governance within science have emerged. One of these is the proposal to include practitioners, people involved in the application of scientific discoveries and with experience in this domain, as part of scientific self-governance institutions (Fuller 2006: 162). They could make other researchers aware of any eventual unintended consequences of a given project already during its preparation and implementation phase. These types of changes however, given the strength of external market mechanisms directing technoscience today, do not seem sufficient. Other means of controlling the direction of research need to be developed alongside these self-governance mechanisms. In such a case, institutional means of monitoring become indispensable (Fuller 2006: 163). Let us therefore stress: the reaction to the problem of risk generated by technoscience and to the potential destabilisation of the collective cannot depend on the (necessarily) slow, bottom-up reform of world-views (even the world-views of experts themselves).

As the previous paragraphs demonstrated, like Latour, Beck claims that we dispose today, at the very least, of the rudiments of international institutions, which shall in the future guarantee the desired level of responsiveness to the problem of risk. These include NGOs, social movements and global media. Beck expresses the hope that new, transnational negotiation systems, in which decisions are no longer made by experts alone, shall emerge from them. They shall uncover and make transparent the premises on which risk is defined (Beck 2000: 223). However, it must be noted that Beck's programme remains imprecise.

When looking for projects enabling a concrete monitoring of technoscience, we find a series of interesting, already extant institutions created to assess actual examples of technological risk. Institutions evaluating innovations are related to Technology Assessment (TA), which appeared in the 1970s. The TA movement is one of the domains within the study of the future of technology in which the latter is deemed controllable. It employs a method of scenarios and impact assessment, taking into account vested interests and social values (Zacher 1986, 2002: 173). The social evaluation of technology takes place within an interdisciplinary framework (while also assuming that scientists and engineers are not entirely competent in this respect), and also within a global perspective, taking into account the social and ethical dimensions of innovation. As we can therefore see, it fulfils the initial postulates brought forward by Beck and Latour.

And so, from 1972 to September 1995, the Office for Technology Assessment (OTA) was an institution at the US Congress. A similar office has also existed at the Bundestag since 1985. These are examples of institutions created in order to influence parliamentary decisions, mainly by preparing expert reports. However, they can also employ participatory and activising methods when creating these. It is interesting to note that institutions tasked with evaluating technology are federated within a European framework – the European Parliamentary Technology Assessment (EPTA). Created in 1990 under the patronage of the president of the European Parliament, EPTA serves as a consulting body for European parliamentary institutions, while also increasing cooperation between bodies tasked with evaluating the impact of new technologies.[28]

In the opinion of Lech W. Zacher, a Polish sociologist addressing issues of technology, the movement pushing for a social evaluation of technology is decidedly different from previous ways of thinking about the role of technology. It avoids technophobia, antitechnicism and catastrophism by default (Zacher 1986: 193). It is however worth noting (as Wiebe E. Bijker does for instance), that TA was initially characterised by technological determinism. It presupposed the possibility of unequivocally defining the consequences of a given innovation and therefore the chances of effective intervention. It therefore emerged from a completely different context than today's risk-monitoring institutions (Bijker 1993: 129). In most cases, the Technology Assessment was based on unquestioned

---

28 On the Polish side, the Biuro Analiz Sejmowych (Sejm Bureau of Research) is an associate member of EPTA. See http://www.eptanetwork.org/, accessed 11.11.2011. It is also worth noting that the Polskie Towarzystwo Oceny Technologii (Polish Association of Technology Assessment) was created in Warsaw in 2013.

expert knowledge, which, in view of the analysis presented above, can be seen as a major drawback.

When searching for legal and political instruments designed to ensure early prevention of the modern risks generated by science and technology, we come upon the Precautionary Principle. This principle is widely discussed and has many different formulations, as well as stronger and weaker versions. It is also sometimes interpreted as an "approach" rather than a principle. Even when it is precisely defined, it also generates problems and controversies when applied on a practical level. We find the first formulation of the Precautionary Principle in a German environmental protection programme from 1971. The principle was introduced into German law in the 1970s. It also found expression in the environmental programmes of Denmark, Sweden and France.

The first international resolution adopting the Precautionary Principle was the World Charter for Nature, adopted by the UN General Assembly in 1982. The Montreal Protocol, adopted in 1987, also contains the principle. It was later included in the Maastricht Treaty in 1994, as well as EU environmental law (Andorno 2004: 14). The Rio Declaration on Environment and Development from 1992 also contains its worldwide formulation. In its formulation from 1998, contained in the Wingspread Statement on the Precautionary Principle, it states: "When an activity raises threats of harm to human health or the environment, precautionary measures should be taken even if some cause and effect relationships are not fully established scientifically. In this context the proponent of an activity, rather than the public, should bear the burden of proof. The process of applying the Precautionary Principle must be open, informed and democratic and must include potentially affected parties. It must also involve an examination of the full range of alternatives, including no action" (in: Klaassen 2007: 108–109).[29]

Worldwide consensus as to the application of the Precautionary Principle, as well as the references made to it in international courts of law, clearly demonstrate its current status as international law (Andorno 2004: 15). The principle finds a formulation in nearly every international treaty relating to environmental protection. However, views on its applicability as a stable law are still divided.

The Precautionary Principle is an instrument used to limit potential risk when we do not yet possess scientifically proven evidence of a potential threat. It is a very useful tool for blocking innovation when there is a lack of scientific certainty

---

29 The Wingspread Conference on the Precautionary Principle was organised by the Science and Environmental Health Network.

as to risk, which, as we have shown, is often a defining characteristic of modern society. By definition, the principle applies to serious, irreversible threats to the environment or public health. It is clear that we often do not possess the instruments to unequivocally evaluate the seriousness of a given threat, or its irreversibility. Despite this, the principle we are discussing legitimises prudent restraint from action in place of carefree experimentation. It stipulates that experts evaluating the validity of the use of the Precautionary Principle should be autonomous vis-à-vis actors with a vested interest in ignoring it and refraining from taking it into account. The principle also underlines the necessity of maintaining transparency and democratic openness during the process of its application. It also allows for the free participation of all those interested in its implementation. These elements are extremely important since they make it more difficult to pervert this rule, for instance by subverting it into a tool for blocking investment efforts by the competition. However, the Precautionary Principle does not constitute a ready-made algorithm for making difficult political decisions.

Both Beck and Latour approve of the need to use the principle within the context of solutions designed as part of technoscience (Beck 2008; Latour 2002: 33). Beck defines it as the "principle of precaution through prevention", which should be employed instead of the current policy of acting after the fact. The principle states that whenever the introduction of a particular solution indicates that there is a risk, the process should be halted until its harmlessness has been proven. Thus, instead of assuming the *a priori* innocence of a given change, we should take a step back, allowing for the possibility that it may have unintended side effects. What is significant, according to the Precautionary Principle, the costs of proving the harmlessness of an innovation should be borne by the implementing party rather than the potential victims, in accordance with the "polluter pays" rule. The existence of post-modern risk, global and irreversible in nature, justifies this type of procedure. It seems immoral to expose others to risk when taking decisions in their name.

It is therefore hardly surprising that we observe strong opposition to the use of the Precautionary Principle from actors that most often introduce new solutions into the collective through market mechanisms. The main argument used by critics of the principle is that there is no real possibility to test all the possible effects of a particular innovation (Klaassen 2007: 109), while demonstrating the absolute harmlessness of anything is an impossible task. Due to methodological limitations, science is incapable of proving that something doesn't exist. All that can be demonstrated is that certain effects have not been observed at a certain point in time. However, as Roberto Andorno notes, the Precautionary Principle does not force the initiators of a given change to demonstrate that the risks relating to it do not appear at all, since that is impossible (Andorno 2004: 19). They should only bear the costs of assessing the risk initially identified by opponents, demonstrating that they have carried out all the analyses available at a given time and have defined the nature and reach of potential threats generated by the innovation in question.

Opponents of the Precautionary Principle also draw attention to the fact that sometimes, for various reasons; the risks of a particular innovation are worth bearing. For instance, houses built in the USA using asbestos, even when its toxicity was already known, protected the health and lives of their inhabitants from the effects of fires (Perelman 2005: 136–137). It has therefore been suggested to reverse the Precautionary Principle as a principle, since no innovation-blocking rule should be imposed unless unequivocal proof can be provided that a particular innovation should be blocked (Perelman 2005: 137).[30] It would seem that this postulates a return to the situation from before when the principle was formulated, when the victims of threats had to finance the procedures and analyses demonstrating the nefariousness of a given threat, legitimising the need for regulating it through costly judicial procedures.

Latour directly discusses commonplace critiques of the Precautionary Principle which state that it blocks all human activity and innovation. In his opinion, the point is precisely that this principle should modify our perception of *each act*, allowing to transparently attribute the unintended consequences of innovations to their initiators (Latour 2009a: 69). We cannot allow ourselves the comfort of free experimentation any longer, since the costs of this can prove far higher than what we can bear. Undesired side effects always accompany our interventions, however within a risk society, we should follow their trajectory and discuss them openly. They are the natural costs of our actions. From the moment the interventions of technoscience and industry intensified, entwining the fate of mankind with that of numerous non-human actors, our control of and mastery over nature must above all be expressed through our connection with it, as well as through care and responsibility. What type of responsibility are we interested in? What type of care are we discussing here? I shall attempt to answer these questions in the following section.

---

30 A modification of the Precautionary Principle has also been posited, combining it with decisions based on current knowledge (Welsh 2006: 163 and following). In such a case, it would not entirely block innovation, allowing, in certain circumstances, to learn how to develop technologies safely.

## The macro-ethics of global responsibility[31]

Examining the problem of the undesirable consequences of the success of technoscience as the problem of the *influence* of science and technology on society is not the right approach, since when making this analytical decision we presuppose the independence of both technoscience and society. Meanwhile, the presupposition of a dichotomy between things and people, the social and the technological or objective world, does not allow one to perceive the deep mutual interrelationship between both domains. Things and technologies have played an important role in the history of civilisation. On the other hand, what we call society is not easy to separate from things. Often, the very appearance of a gadget or innovation means that it has started to live a life of its own, transforming the collective, and that this process cannot be stopped, and even less so reversed.

Reflection on the role of discoveries and innovations should precede their appearance, and it should question the very probity of carrying out particular research. It cannot take place after the fact. This in turn means that in the era of risk, the scientific belief in the unproblematic, positive role of discovering (what is new and unknown) in laboratories is simply politically dangerous. The Precautionary Principle should therefore already be implemented at the level of basic research.

The political problem of risk derives from the fact that when the products (and discoveries) developed as part of technoscience encounter no barriers, they reach the market too quickly (Stankiewicz 2004: 154). This mechanism cannot be blocked without tampering with the axiological foundations of the global economic order and the main, modernistic rules of a free market economy, such as the absolute autonomy of scientific research and economic institutions, the priority of profits and the ever increasing speed of development. It is however worth noting that market values are not limited to profit, but also include expansionary logic as well as the requirement to continuously increase the quality of life, which is extremely difficult to question at the level of the collective consciousness.

If we do accept the policy of acting after the fact, by giving a voice to ethicists, sociologists, philosophers, cultural studies researchers or ordinary citizens only after the introduction of innovations and new hybrids into the collective domain, we also lay the foundations of our own, systemically planned helplessness. In most cases it is already too late. The only remaining option is to adjust our axiological systems and practices to what has already taken place. The question of global responsibility for innovation is not even raised.

---

31  For a discussion of the political role of laboratories, participatory turn and the macro-ethics of global responsibility see Bińczyk 2013.

Dieter Birnbacher, a German technology researcher, suggests that when introducing innovations we should be guided by the criterion of reversibility, only permitting those undertakings which can be blocked if necessary, or even reversed, without serious economic or environmental consequences. He writes: "one should not establish accomplished facts. This means that before making a decision as to the development of a given technology, one should – as thoroughly as possible and taking into account alternative positions – research and assess the consequences of these technologies for people, society and the environment" (Birnbacher 1995: 677). The process of thinking about the consequences should precede any investment procedure. It is however clear that the responsibility ensuing from any side effects due to the introduction of technological innovations is incredibly hard to attribute accurately: "No one can be blamed for the consequences of his actions or inaction if these are not foreseeable, avoidable, or if avoiding them would require a heroic effort" (Birnbacher 1995: 676). The question of the adequate attribution of responsibility therefore needs to be examined and regulated.

As it would seem, in an era when the destabilisation of the collective as a whole has become possible, not only should ethics not be left to its own devices, but the problem of the political and ethical reactions to the unintended consequences of technoscientific interventions should also not be approached from an overly narrow individual perspective. An appeal for a necessary ethical progress, a change in the positions of individual scholars or even citizens is not sufficient, since it does not translate into concrete institutional and systemic changes. Informing, enlightening or sensitising people to the social responsibility of scholars and ordinary citizens when it comes to risk is much too little. The transformation of ethical positions, even on a large scale, can in fact lead to no changes at all in the global mechanisms regulating the functioning of technoscience in the modern world. Moreover, this naive postulate of transforming individual positions can in fact prove harmful by paralysing political action. It thus follows that we need "a different politics, not a different psychology" (Latour 2004: 257, footnote 29).

The necessity of devising a macro-ethics of shared responsibility, which must extend beyond conventional morality, was underlined by the German philosopher Karl-Otto Apel (Apel 1995). The invention of nuclear weapons, the creation of global interactions as part of a worldwide market, but also the ecological problems of the 20th century, require examining the issue of responsibility for collective action.[32] Piet Strydom, a researcher on risk, has also written about the

---

32 It is worth noting that Apel, the creator of the transcendental-pragmatic theory of communication, designed his macro-ethics as universal, rationally justified, based on

macro-ethics of global responsibility (Strydom 2002: 129). However, we must remain aware that in all certainty, these ethics shall not be created by our institutions, which mainly function according to the market logic of profitability. Industry limits itself to employing long-term policies of adapting to the changes taking place currently, and for instance, is already planning to reap future profits from a society dealing with the effects of climate destabilisation. Institutions to guarantee an effective global policy of responsibility will therefore probably not be initiated by the most prominent actors within the current economic order. Yet despite all this, an ever increasing number of intellectuals are building macro-ethical utopias. Let us take a closer look at these.

Let us start with the project of an ethics of responsibility devised by the German philosopher Hans Jonas, whose book *Das Prinzip Verantwortung: Versuch einer Ethik für die technologische Zivilisation* (Jonas 1996) was first published in German in 1979. As we shall see, Jonas's position anticipates many of the theses included in this study. First of all, his book states that modern technology transforms the basic parameters of human action (Jonas 1996: 21), expanding its reach and scope of intervention. As a result of human activity, the difference between what is created and what is natural becomes blurred. Man can technologically (genetically) transform himself. What is more, "modern megatechnology" is capable of ending planetary life and humankind itself (cf. Levy 2002: 138). Jonas also highlights the frequent irreversibility of technological interventions (Jonas 1996: 72). The rate of change introduced by technology renders a balanced auto-correction impossible.

We have never before been faced with a situation of this type. According to the author of *Das Prinzip Verantwortung*, its specificity requires a transformation of the very foundations of ethics. Previous theories of morality were characterised by the following assumptions: 1) treating the relationship of man with the non-human realm (*techne*) as completely neutral ethically, 2) anthropocentrism – ethics used to deal with man's relationship with man, 3) "man" and his condition were seen as fundamentally constant, impervious to change by *techne*, 4) ethical principles and values were focused on the "here and now", situated within the shared present of acting entities (Jonas 1985: 26–27). Until now, ethics only considered non-cumulative effects, while currently "effects keep adding themselves to one another, with the result that the situation for later subjects and their choices of action will be progressively different from that of the initial agent"

---

the non-contingent preliminary assumptions of argumentative discourse. His views remain quite distant from most of the positions enumerated and accepted in this book.

(Jonas 1996: 31). Furthermore, being moral and making moral decisions did not previously require expert knowledge. In Jonas's view, *all* these assumptions are presently void.

The uncertainty characterising the modern era legitimises a position of responsible restraint. When the domains impacted by risk expand, so do those of responsibility. Jonas writes about preventative responsibility, related to the possible consequences of our actions (*ex post* responsibility concerns what we have already done). As we have seen, the Precautionary Principle is precisely such an instrument allowing to implement preventative responsibility.

Jonas reformulates Immanuel Kant's categorical imperative in such a manner that it covers both the requirement of maintaining the integrity of mankind and the integrity of life. Jonas's responsibility principle states: "Do not compromise the conditions for an indefinite continuation of humanity on earth" (Jonas 1996: 38). Similarly, Jonas proposes to extend responsibility to future generations as well. The ethics of responsibility are not based on reciprocity – I cannot ask what the future can do for me, since future generations cannot repay us in kind.

What is interesting is that Jonas's ethics of responsibility concern both the private and the public sphere. Our current situation requires a profound redefinition not only of ethics but also of politics. The author of *Das Prinzip Verantwortung* writes: "morality must invade the realm of making, from which it had formerly stayed aloof, and must do so in the form of public policy" (Jonas 1996: 35). As he notes, modern governments consider current issues, "But the future is not represented (...). The nonexistent has no lobby, and the unborn are powerless" (Jonas 1996: 22). This situation has to be altered (in a democratic manner). Unfortunately, Jonas gives us no precisions as to how this should be done. The German philosopher merely notes the necessity of introducing an "heuristics of fear", which implies making a public habit of always imagining the worst possible consequences of our actions first (even when the best possible intentions stand behind them).

In sum, Jonas's view is one of the most interesting and fleshed-out positions within the macro-ethics of global responsibility. His diagnosis, already partly formulated in the 1950s (cf. Jonas 1996: 18), seems to be fundamentally correct, though it is unfortunately not developed enough. The author of *Das Prinzip Verantwortung* does not employ the notion of risk, but he does highlight the far-reaching and irreversible consequences of modern technology, which transforms the very nature of human action, and thereby also the nature of responsibility. He also states that science does not provide any absolute certainties on which a technological civilisation could be based. Jonas ceases to treat technological innovations neutrally, as innocent gadgets, while also speaking of the necessity of abandoning a purely anthropocentric ethics. Writing on politics, Jonas admits taking into account the political representation of non-human actors as well as the rights of future generations. Introducing reflection on preventative responsibility or restraint, he legitimises the reasonableness of introducing legislation such as the Precautionary Principle. Jonas also discusses the fact that, on account of the scope of technological intervention, the human condition can no longer constitute an unchanging starting point for ethics. This assertion forms the basis of the current post-humanism, critical with regard to the assumptions of essentialism.

Unfortunately, Jonas's approach remains largely theoretical and axiomatically anthropocentric (Jonas considers the existence of moral agents capable of responsibility to be of absolute value, which justifies his attempts to maintain the integrity of life on Earth). In *Das Prinzip Verantwortung*, he devotes most of his attention to devising metaphysical justifications for his position. Let us note that Jonas's imperative of responsibility forms the basis of many ecological and bio-ethical positions. It remains strongly linked to many important ideas of the late 20th century, such as sustainable development and collective responsibility (towards other living organisms and future generations).

The ideas of sustainable development and collective responsibility are interesting modern axiological reactions to the problem of the fair distribution of risk. They mostly concern ecological risk and global social inequality, while also being an attempt to create a global macro-ethics of responsibility. Sustainable development is a postulate, or idea, of a use of natural resources that does not limit the ability of future generations to satisfy their own needs. This idea presupposes that the implementation of ecological postulates should be harmonised with economic development and the ideal of social justice. The postulate of sustainable development appeared at a UN session on the environment in 1972. It was related to discussions concerning limits to growth in the wake of the Club of Rome report. Sustainable development, as well as collective responsibility, are also mentioned in the *Our Common Future* report of the World Commission on Environment and Development, published in 1987. The Brundtland Commission defines sustainable development as "development that meets the needs of the present without compromising the ability of future generations to meet their own needs" (Brundtland 1987: 326). The postulates cited above were made the topic of international discussion during the UN World Summit conference in Rio de Janeiro in 1992. This summit resulted in the Rio Declaration on Environment and Development and the Agenda 21 action plan. The debate on sustainable development signifies the acceptance of the fact that ecological stability, social justice, participatory democracy and economic growth must always be considered as interdependent.

It would seem that the ideas of sustainable development and collective responsibility are an important attempt to accentuate the importance of the rights of future generations, and a political, transnational attempt to articulate the interests of non-human actors (in this case, endangered species, environmental characteristics, ecosystems). Their appearance in the global political arena has its own historical roots and is gradually taking on an institutionalised form. Unfortunately, we do not have room here for a detailed analysis of the evolution of the idea of sustainable development, or its implementation in different countries. We should however note that the idea of sustainable development only provides imprecise guidelines, necessarily subject to many diverse interpretations. Some even view these concepts as an analytically flawed and overly abstract set of slogans (Giddens 2010: 71–72). Moreover, sustainable development does not question the need for further economic growth, including the hope that businesses will provide new technologies that are salutary within the context of dwindling planetary resources. Criticism of this concept also includes the objection that it is an instrument used to impose developmental barriers on developing countries in the name of the aesthetic preferences of the elites in developed ones. I believe that we should approach each individual attempt at implementing the principles of sustainable development critically and carefully.

Certainly ethical norms and theories should be adapted in such a way as to be able to deal with the existence of complex global links between heterogeneous elements. One of the attempts at reaching this type of goal is the "long-range ethics" of the Polish philosopher Krzysztof Abriszewski. In many ways, his long-range ethics are similar to Jonas's ethics of responsibility. Appealing to the actor-network theory, Abriszewski writes that the modern links between human and non-human actors are becoming increasingly more extensive: "morality is delegated to the thousands of intermediaries connecting us" (Abriszewski 2007: 284). Between the shirt I buy and the Chinese seamstress who sewed it, we find a whole series of heterogeneous elements. We should become sensitive to these far-reaching relationships, those linking humans to non-humans, as well as to their consequences, including ethical ones. Artefacts, medical procedures and technologies co-create morality.

The need for a special redefinition of ethics is also underscored in the book *Chasing Technoscience*. We should extend the notion of responsibility, since it is currently not limited to individuals taking actual decisions (Ihde, Selinger 2003: 194). We can observe effects that extend beyond the intentions contained in the "here and now". People are always elements within wider, complex, socio-technological networks of links. When conceptualising the issue of responsibility in modern society, we cannot avoid relational thinking. It is precisely in this place that the concept of a network can find its ideal application. Global links, the surprising, distant (both in time and place) effects of interactions, the mutual imbrication of various types of consequences, all this implies that we cannot situate responsibility solely within the intentional domain of individual (human) actors. Often, our decisions have far-reaching consequences, both dangerous ones resulting in risk and beneficial ones, which is precisely why we have to formulate a long-range ethics.

As it would seem, globalisation has already revealed new types of solidarity and ethical sensibility. For instance, thanks to the internet and television channels such as CNN, we increasingly treat distant problems as our own. We identify with the problems of people we will never actually meet. Another example is the possibility of providing so-called micro-loans, which reach interested parties directly, at a distance.[33] Within modern society, thanks to telecommunications infrastructures, the spontaneous moral impulses which arise in face to face contact could develop into a global moral imagination. One of the ways of implementing a long-range ethics could be termed the idea of informed consumption, including the popularisation of fair-trade "brands". The necessary condition for the proper functioning of informed consumption is the availability of information on the provenance, production and importation of specific products.

In order to systematise current solutions, let us now list the most important parameters of a macro-ethics of global responsibility:

1) it is above all an ethics that includes the complex relationships between heterogeneous entities which form the collective, and not an ethics of face-to-face reactions in the "here and now",
2) it extends the scope of the notion of responsibility (including unborn generations and their needs, life on Earth, other species etc.),
3) in the same manner, it is non-anthropocentric (it encompasses relationships between humans and non-human actors),
4) its dynamics cannot consist only in a post factum adaptation to the interventions of technoscience,
5) the macro-ethics of global responsibility must be systematically institutionalised using technologies and non-human actors (it cannot be reduced to postulates for modifying the positions and world-views of individuals),
6) it demands a re-examination of the axiomatic foundations of the global economic order (for instance, by replacing the notion of progress with that of sustainable development).

---

33 For instance, see http://www.kiva.org/.

## Conclusion

Obviously, the postulates of monitoring the expansion of the collective and regulating technoscience are set within a defined set of assumptions. Some of these are axiological – the management of risk and laboratory interventions is performed in order to ensure a certain future state, valued positively. The current nature of risk is always co-defined by a tacitly presupposed, particular vision of the common good. However, what is somehow more important, since the results of this monitoring must appear to be predictable, the controllability of social, technological, environmental and cultural processes is also assumed. Because of this, it is worth re-emphasising that both Beck and Latour (as well as the other authors cited in this work) all formulate postulates which do not as a matter of fact undermine the ideals of western democracy or even a specifically understood instrumental rationality. After all, they desire to place the further development of science (as well as technology) under rational and democratic control (cf. Harding 2007: 33). But is rational and democratic control of the future of the collective at all possible?

Even in the work of Beck, Giddens and Lash we find expressions of scepticism as to this type of endeavour. They write: "New areas of unpredictability are created quite often by the very attempts that seek to control them", "The more we try to colonize the future, the more it is likely to spring surprises upon us" (Beck, Giddens, Lash 2009: 9, 82). This probably means that these scholars do not in fact believe in a satisfactory, political control of the future. The current political framework is seen here as a domain that is not entirely predictable. Interventions intended to increase controllability in fact also increase complexity, and so paradoxically make monitoring even harder. In fact, the modern risk found in Beck's work displays an unpredictable, surprising nature – it escapes rational management. We can also see that, as a result of the emergence of global associations between ontologically diverse elements, we are trapped in the midst of the unforeseeable side effects of our own interventions. This is why researchers studying modern, mutually overlapping technological structures underline the fact that they are chaotic and immeasurable. Given this, we can observe a dilemma, which reappears at many points in this narrative: how do we monitor the expansion of the collective at a time when the collective itself has become unpredictable? How do we measure and manage risk which has become immune to measurement and management?

According to Niklas Luhmann, a German sociologist, every social subsystem (science, the economy, politics) is able to perceive risk only according to its own internal differentiations. These perceptions are incommensurable, while ecological issues are perceived by the individual subsystems solely as perturbations

(Luhmann 1991: 96–97). In Luhmann's view, a new risk policy (monitoring, earlier debates, mutual controls, institutionalised means of opposition etc.) will only increase the complexity and risk already existing at the decision-making level. It will not facilitate proper action and will also not eliminate the difficulty of unequivocally differentiating between the positive and negative consequences of the changes introduced (cf. Strydom 2002: 68). What is more "the risk-eliminating risk remains a risk" (Luhmann 1991: 30) – our interference, meant to protect us against a given danger, can in fact provoke further unintended consequences.

What this implies is that it is better to let matters run their own course. Luhmann calls his position regarding risk "scepticism". Instead of solving the problems of a risk society, he proposes to accept them. The social sciences should also accept this problem. Each social subsystem defines risk according to its own criteria and deals with it.

It would seem that Luhmann's position regarding the problem of eventual political reactions to risk is quite convenient, but also overly optimistic and probably insufficient. It implies an attitude of arbitrarily absolved passivity, or in the best case, the transposition of the problem of monitoring technoscience to the local level of the individual subsystems within the collective.

However, as the political philosopher Alasdair MacIntyre notes, any scientific diagnosis of modern society and prevision of future social phenomena can in fact prove impossible (MacIntyre 1996: 170–206). In his renowned work *After Virtue. A Study in Moral Theory*, we read "Our social order is in a very literal sense out of our, and indeed anyone's, control" (MacIntyre 2013: 124). This derives from the nature of social life itself, which is based on free decisions. If MacIntyre is right, we return yet again to a comfortable, arbitrarily absolved passivity. We leave further expansion of the collective to market logic and its own course.

Let us underline for the final time that the acceptance of the possibility of achieving relatively correct prognoses (and recommendations) in the social sciences does not stray far from the technocratic belief which privileges the bureaucratic professionalism of the expert-manager. Both elements are tightly linked together. This implies that it is always judicious to treat the recommendations of experts with a certain amount of caution, even when those experts are humanities scholars like Beck or Latour or other exponents of science and technology studies, imposing political solutions legitimised by a particular version of the "terror of the future".

Despite everything, I believe that the restrictions and difficulties mentioned here do not legitimise accepting the passivity of the humanities (and of all of mankind) when it comes to the problem of monitoring our own future. The helplessness of scholars when it comes to effectively predicting the future, imposes an even greater responsibility. The humanities and social sciences (despite their far-reaching fragmentation and specialisation) cannot, and should not, dodge the requirement of devising alternate scenarios. Their tasks should include simulating current tendencies and the consequences of changes, and examining our goals as well as the directions we are heading in.

Let us therefore examine the goals which, I believe, we should be aiming for. While attempting a synthesis of all the proposals outlined above, I shall list the most important philosophical and political recommendations on the question of the unintended consequences of the success of technoscience.

1) Technoscience and its laboratories should be subjected to democratic monitoring. The Precautionary Principle should be employed reasonably and transparently. The very point of financing and carrying out particular research should be discussed before any research is actually done. The same should hold for the directions of development of technoscience.

2) The most significant mechanisms technoscience is subject to today should be exposed and subjected to public debate. We should speak openly about the consequences of the commercialisation of science, its links to industry and politics and its conflicts of interest. Currently, the most vital decisions concerning the future direction of research are taken within the realm of sub-politics, at the intersection of global consortia (biotechnology, pharmaceutics, defence industry), privatised laboratories, superpowers and political lobbies. The role of pure basic research has become negligible, while science is taking on a post-academic character. Because of scientistic ideology, these processes are either ignored or disregarded as pathologies. This needs to be challenged.

3) We should strive to create institutions enabling open, public debate on the innovations (including risks and controversies) faced by the collective. We need to devise effective practical solutions that guarantee an open, honest, conclusive and civic debate on the direction of research and proposed innovations. The point of these debates shall not be to reveal an immanent, final vision of the public good, but a temporary, underdetermined yet robust adaptation within the space of the articulated points of view, values and negotiated interests.

4) Rejection of the paternalism of natural science experts and of the state. Not only experts, government and industry representatives, but also sociologists, cultural scholars, ethicists, economists, credible representatives of non-human actors, advocates of the rights of future generations, spokespeople for groups directly interested in the consequences of innovation and even ordinary citizens should be included in the debate on the future of the collective and on the role

of innovation. The question of the complex status of expert knowledge should be approached openly.

5) The relationship between technoscience and society should not be analysed in terms of the "influence" of one independent domain on another. Artefacts and technologies have always been a part of the human world and continue to be so. Technoscience interferes deeply in the structure of the world we live in (the social, ecological, economical and normative domains are all intertwined here). The thoughtless incorporation of technological innovations and scientific discoveries into the collective often results in deep, surprising consequences in sometimes entirely unrelated domains. Within the context of the symptoms of change outlined above, it seems appropriate to replace the term "society" with the notion of "collective" proposed by Latour. The collective is formed by people and non-humans (defined as natural or technological). These entities are all interdependent. Non-human actors should be included in the political domain (not only by discussing them and through their linguistic representation), but also by addressing directly the problem of our responsibility towards them (as well as that of our responsibility towards unborn generations). This signifies abandoning anthropocentrism within politics and ethics.

6) Intellectuals should thoughtfully problematise the main assumptions of modernism. Scientific knowledge, laboratory activities and the generation of new discoveries should no longer be perceived as an unproblematic good. Technological innovations should not be presented as unquestioned tools of progress or innocent, isolated gadgets. The image of technoscience should be profoundly transformed – it is a domain of practices infected by risk, inherently uncertain, requiring constant tinkering and often involving the possibility of error. Moreover, progress does not have to be perceived as inevitable in its current state. Some "goods" should not be produced at all (due to pollution, risk or the simple triviality they generate). Values such as profit, growth, the constant raising of living standards and unconditional freedom of investment endanger the future stability of the world.

7) The specific matrix of "organised irresponsibility", in which no groups bear responsibility: scientists merely discover facts, industry generates profits, politicians fight for power and citizens remain passive, unable to believe in what indeed is a very real possibility of influencing the shape of the world we live in – this needs to be challenged or even totally dismantled. According to Zacher, the reason for organised irresponsibility is the "functional differentiation of society", divided into the autopoietic operational systems of science, economy and state, governing themselves through their own internal logic (Zacher 1994: 28). The dismantling of this matrix would require the implementation of mechanisms that guarantee the attribution of responsibility within individual systems in such

a way that, for instance, the voices of ethicists or sociologists are not treated as "ambient noise" within the economy or the state. The internal logic of functioning of individual subsystems can therefore no longer be examined in isolation, but rather as part of the challenges of a global ethics of responsibility.

8) We must create global legislation as well as transnational institutions that permit the systemic monitoring of technoscience.[34] We should strengthen processes enabling the creation of global, independent media and global civic movements. If technoscience finds itself in a post-academic phase, then monitoring it means monitoring market interdependencies and limiting the freedom of the major actors of economic globalisation. A proper identification of the central sources of causality, the decisional domains within sub-politics and the global economic structure are the most important problem facing modern economists, lawyers, politicians, advisers and decision-makers.

As implied above, the monitoring of technoscience should take place over many, mutually supportive and overlapping levels of action. Firstly, this would be the level of global institutions and transnational legislation (international agreements, regulations, legislation, the Precautionary Principle etc.). The second level would be the domain of state authority, parliamentary decisions, legal restrictions within individual countries (technological education, Technology Assessment institutions, citizen boards, science shops). Thirdly, the point is to monitor the future of the collective from within technoscience itself, and therefore from within industry, with which it is completely interwoven. This implies including interdisciplinary reflection on the future of the collective in the process of defining new areas of research (and future profits) and in the process of making scientific discoveries or constructing technological innovations. Such results may be attained through the creation of scientific councils taking into account a large spectrum of voices and values, or through the introduction of agreements between industry, NGOs and ecological organisations). Finally, we must underline the need to create a platform for a conclusive, effective and open public debate on the role of technoscience, including the participation of laypeople and diverse interest groups. The points of view of groups impacted by the consequences of a given intervention would need to be satisfactorily articulated. A democracy, in which a rational debate of the type described above would be

---

34 Francis Fukuyama, writing on the consequences of the biotechnological revolution, describes these types of solutions (Fukuyama 2008: 245 and following). He underlines the fact that within the context of the monitoring of domains such as energy, nuclear, chemical and biological weapons, human organs and neuropharmacological substances, international regulations have proved effective.

possible, should at the same time guarantee a specific informational context. The point is to reveal the actors and interests defining technoscience and industry, as well as risk. Furthermore, public opinion should be allowed to ask questions about the legitimacy and axiological background of specific actions, in addition to being informed about the potential side effects of discoveries and the far-reaching consequences of innovation.

The recommendations presented above will probably not fully satisfy the reader searching for ready-made intellectual, political or philosophical solutions. However, I do hope that, at least in some places, my argumentation shows that the unintended effects of the success of technoscience can (and should) be discussed in a completely new and non-standard way. This is a matter of perceiving technoscience and its role in a way that challenges routinely accepted beliefs on society, the dynamics of social change, effectiveness, laboratories, infrastructures, progress and the need for growth. It is precisely these assumptions which obstruct the thinking on risk and prevent an adequate recognition of the role of technoscience. There are many mental mechanisms of rationalising passivity: a disdain for utopias, the discounting of future threats, which we do not want to consider seriously, the paternalistic rejection of values articulated by laypeople, the establishment of limit values for risk, the belief in the human ability to fix all damages in the future. The Enlightenment vision of progress places severe restrictions on our imagination, making it difficult for us to imagine alternatives (apart from the inevitable pursuit of further growth, discoveries and innovation).

As it seems, within the realm of sociology of risk and science and technology studies we are faced with a particular conceptual revolution, taking place in front of our eyes. It establishes a new vocabulary for our thinking about the civilisational role of technoscience. Up till now, we have been silently aware of the existence of the undesired consequences of our endeavours. Yet in the era of modern risk, the need to react is pressing. The problem is not only that of the protection and comfort of the current collective, closest to us, but also of preserving future, still possible and unknown worlds.

## Literature

Abriszewski, Krzysztof. 2007. *Interpretacja i etyka dalekiego zasięgu*. In: Adam F. Kola, Andrzej Szahaj (ed.). *Filozofia i etyka interpretacji*. Kraków: Universitas, 277–288.

Afeltowicz, Łukasz. 2011. *Laboratoria w działaniu. Innowacja technologiczna w świetle antropologii nauki*. Warszawa: Oficyna Naukowa.

Andorno, Roberto. 2004. *The Precautionary Principle: A New Legal Standard for a Technological Age*. "Journal of International Biotechnology Law" 1: 11–19.

Apel, Karl-Otto. 1995. *Problem uniwersalnej makroetyki współodpowiedzialności*. In: Tadeusz Buksiński (ed.). *Wspólnotowość wobec wyzwań liberalizmu*. Poznań: Wydawnictwo Naukowe IF UAM, trans. Izabela Ferenc, 33–50.

Bauman, Zygmunt. 2007. *Szanse etyki w zglobalizowanym świecie*. Kraków: Wydawnictwo Znak, trans. Jacek Konieczny.

*Bayh-Dole Act*. 1980. *Public Law 96–517, 96th Congress, 2nd Session* (12 December).

Beck, Ulrich. 2000. *Risk Society Revisited: Theory, Politics and Research Programmes*. In: Barbara Adam, Ulrich Beck, Joost van Loon. *The Risk Society and Beyond. Critical Issues for Social Theory*. London, Thousand Oaks, New Delhi: SAGE Publications, 211–229.

Beck, Ulrich. 2002. *Społeczeństwo ryzyka. W drodze do innej nowoczesności*. Warszawa: Wydawnictwo Naukowe Scholar, trans. Stanisław Cieśla.

Beck, Ulrich. 2005. *Władza i przeciwwładza w epoce globalnej. Nowa ekonomia polityki światowej*. Warszawa: Wydawnictwo Naukowe Scholar, trans. Jerzy Łoziński.

Beck, Ulrich. 2008. *Ekologiczny atom?* „Dziennik. Europa", 26 July, No 30 (225): 12.

Beck, Ulrich. 2009. *Na ile realna jest katastrofa klimatu?* In: *Ekologia. Przewodnik Krytyki politycznej*. Warszawa: Wydawnictwo Krytyki Politycznej, trans. Michał Sutowski, 76–117.

Beck, Ulrich, Anthony Giddens, Scott Lash. 2009. *Modernizacja refleksyjna. Polityka, tradycja i estetyka w porządku społecznym nowoczesności*. Warszawa: Wydawnictwo Naukowe PWN, trans. Jacek Konieczny.

Bhagwati, Jagdish. 2004. *In Defense of Globalization*. New York: Oxford University Press.

Bijker, Wiebe E. 1993. *Do Not Despair: There Is Life after Constructivism*. „Science, Technology, & Human Values", (Vol. 18) 1: 113–138.

Bińczyk, Ewa. 2012. *Technonauka w społeczeństwie ryzyka. Filozofia wobec niepożądanych następstw praktycznego sukcesu nauki*. Toruń: Wydawnictwo Naukowe UMK.

Bińczyk, Ewa. 2013. *Thinking about the Role of Laboratories Today. Anti-essentialism, Macro-ethics and Participatory Turn*. In: Hajo Greif, Martin Gerhard Weiss (ed.), *Ethics, Society, Politics*, Proceedings of the 35th International Wittgenstein Symposium. Berlin, Boston: De Gruyter Ontos, 519–530.

Bińczyk, Ewa. 2013a. *(Post)constructivism on Technoscience*, „Avant", Vol. IV, 1: 317–338.

Birnbacher, Dieter. 1995. *Technika*. In: Ekkehard Martens, Herbert Schnädelbach (ed.). *Filozofia. Podstawowe pytania*. Warszawa: Wiedza Powszechna, trans. Krystyna Krzemieniowa, 647–682.

Böhme, Gernot. 1998. *Antropologia filozoficzna. Ujęcie pragmatyczne*. Warszawa: Wydawnictwo IFiS PAN, trans. Piotr Romański.

Braun, Kathrin, Cordula Kropp. 2010. *Beyond Speaking Truth? Institutional Responses to Uncertainty in Scientific Governance*. "Science, Technology, & Human Values", Vol. 35, No 6: 771–782.

Brundtlandt, Gro Harlem. 1987. *Our Common Future. Report of the World Commission on Environment and Development*. Oxford: Oxford University Press.

Bucchi, Massimiano. 2004. *Science in Society. An Introduction to Social Studies of Science*. London, New York: Routledge, trans. Adrian Belton.

Callon, Michel, Yannick Barthe, Pierre Lascoumes. 2009. *Acting in an Uncertain World. An Essay on Technical Democracy*, trans. Graham Burchell. The MIT Press.

Durant, Darrin. 2009. *Accounting for Expertise: Wynne and the Autonomy of the Lay Public Actor*. "Public Understanding of Science" 17: 5–20.

Durant, Darrin. 2010. *Public Participation in the Making of Science Policy*. "Perspectives on Science", Vol. 18, 2: 189–225.

Durant, John R., Geoffrey A. Evans, Geoffrey P. Thomas. 1992. *Public Understanding of Science in Britain: the Role of Medicine in the Popular Representation of Science*. "Public Understanding of Science" 1: 161–182.

Dybel, Paweł, Szymon Wróbel. 2008. *Granice polityczności. Od polityki emancypacji do polityki życia*. Warszawa: Fundacja Aletheia.

Epstein, Steven. 1995. *The Construction of Lay Expertise: AIDS Activism and the Forging of Credibility in the Reform of Clinical Trials*. "Science, Technology and Human Values" 15: 494–504.

Fishkin, James. 1991. *Democracy and Deliberation: New Directions for Democratic Reform*. New Haven: Yale University Press.

Fukuyama, Francis. 2008. *Koniec człowieka. Konsekwencje rewolucji biotechnologicznej*. Kraków: Wydawnictwo Znak, trans. Bartłomiej Pietrzyk.

Fuller, Steve. 2006. *The Philosophy of Science and Technology Studies*. New York, London: Routledge, Taylor&Francis Group.

Gibbons, Michael, Carmille Limoges, Helga Nowotny, Simon Schwartzmann, Peter Scott, Martin Trow. 1994. *The New Production of Knowledge: the Dynamics of Science and Research in Contemporary Societies*. London: Sage.

Giddens, Anthony. 2010. *Klimatyczna katastrofa*. Warszawa: Prószyński i S-ka, trans. Małgorzata Głowacka-Grajper.

Godin, Benoît. 1998. *Writing Performative History: The New New Atlantis?* "Social Studies of Science", Vol. 28, 3: 465–483.

Harding, Sandra. 2007. *Modernity, Science, and Democracy*. „Social Philosophy Today". *Science, Technology, and Social Justice*, John R. Rowan (ed.). Charlottesville, Virginia: Philosophy Documentation Center, Vol. 22: 17–42.

Held, David. 2010. *Modele demokracji*. Kraków: Wydawnictwo Uniwersytetu Jagiellońskiego, trans. Wojciech Nowicki.

Hughes, Thomas P. 2005. *Human-Built World. How to Think about Technology and Culture*. Chicago, London: University of Chicago Press.

Ihde, Don, Evan Selinger (ed.). 2003. *Chasing Technoscience. Matrix for Materiality*. Bloomington, Indianapolis: Indiana University Press.

Irwin, Alan, Brian Wynne. 1996. *Misunderstanding Science? The Public Reconstruction of Science and Technology*. Cambridge, UK: Cambridge University Press.

Jasanoff, Sheila. 1995. *Science at the Bar: Law, Science, and Technology in America*. Cambridge, MA & London: Harvard University Press.

Jonas, Hans. 1996. *Zasada odpowiedzialności. Etyka dla cywilizacji technologicznej*. Kraków: Wydawnictwo Platan, trans. Marek Klimowicz.

Jones, Mark Peter. 2009. *Entrepreneurial Science: The Rules of the Game*. "Social Studies of Science", Vol. 39, 6: 821–851.

Kerr, Anne, Sarah Cunningham-Burley, Richard Tutton. 2007. *Shifting Subject Positions. Experts and Lay People in Public Dialogue*. "Social Studies of Science", Vol. 37, 3: 385–411.

Klaassen, Johann A. 2007. *Contemporary Biotechnology and the New „Green Revolution": Feeding the World with „Frankenfoods"?* „Social Philosophy Today". *Science, Technology, and Social Justice*, John R. Rowan (ed.). Charlottesville, Virginia: Philosophy Documentation Center, Vol. 22: 103–126.

Klein, Naomi. 2008. *Doktryna szoku. Jak współczesny kapitalizm wykorzystuje klęski żywiołowe i kryzysy społeczne*. Warszawa: Warszawskie Wydawnictwo Literackie MUZA SA, trans. Hanna Jankowska, Tomasz Krzyżanowski, Katarzyna Makaruk, Michał Penkala.

Krimsky, Sheldon. 2006. *Nauka skorumpowana? O niejasnych związkach nauki i biznesu*. Warszawa: PIW, trans. Beata Biały.

Król, Marcin. 2008. *Filozofia polityczna*. Kraków: Wydawnictwo Znak.

Kurczewska, Joanna. 1997. *Technokraci i ich świat społeczny*. Warszawa: Wydawnictwo IFiS PAN.

Kusch, Martin. 2007. *Towards a Political Philosophy of Risk: Experts and Publics in Deliberative Democracy*. In: Tim Lewens (ed.). *Risk: Philosophical Perspectives*. London, New York: Routledge, 131–155.

Kutyła, Julian, Michał Penkala, Sławomir Sierakowski, Michał Sutowski, Agata Szczęśniak. 2009. *Kryzys. Przewodnik Krytyki Politycznej. Przyczyny, analizy, prognozy*. Warszawa: Wydawnictwo Krytyki Politycznej.

Laird, Frank N. 1993. *Participatory Analysis, Democracy, and Technological Decision Making*. "Science, Technology, & Human Values", Vol. 18, 3: 341–361.

Latour, Bruno. 1987. *Science in Action: How to Follow Scientists and Engineers Through Society*. Cambridge, MA: Harvard University Press.

Latour, Bruno. 1991. *Technology is Society Made Durable*. In: John Law (ed.). *A Sociology of Monsters: Essays on Power, Technology and Domination*. London: Routledge, 103–131.

Latour, Bruno. 1993. *We Have Never Been Modern*. New York: Harvester Wheatsheaf, trans. Catherine Porter.

Latour, Bruno. 2002. *War of the Worlds: What about Peace?* Chicago: Prickly Paradigm Press, trans. Charlotte Bigg.

Latour, Bruno. 2003. *The Promises of Constructivism*. In: Don Ihde, Evan Selinger (ed.). *Chasing Technoscience. Matrix for Materiality*. Bloomington, Indianapolis: Indiana University Press, 27–46.

Latour, Bruno. 2004. *Politics of Nature. How to Bring the Sciences into Democracy*. Cambridge, Massachusetts, London: Harvard University Press, przeł. Catherine Porter.

Latour, Bruno. 2009. *Spheres and Networks: Two Ways to Reinterpret Globalization*. „Harvard Design Magazine", Spring/Summer, 138–144, http://www.brunolatour.fr/articles/index.html, 4.10.2009.

Latour, Bruno. 2009a. *Rozwój głupcze! Czyli jak modernizować modernizację*. In: *Ekologia. Przewodnik Krytyki politycznej*. Warszawa: Wydawnictwo Krytyki Politycznej, trans. Barbara Szelewa, 53–75.

Latour, Bruno. 2010. *Splatając na nowo to, co społeczne. Wprowadzenie do teorii aktora-sieci*. Kraków: Universitas, trans. Aleksandra Derra, Krzysztof Abriszewski.

Latour, Bruno. 2010a. *On the Modern Cult of the Factish Gods*. Durham, London: Duke University Press, trans. Catherine Porter, Heather MacLean.

Lave, Rebecca, Philip Mirowski, Samuel Randalls. 2010. *Introduction: STS and Neoliberal Science*. "Social Studies of Science", Vol. 40, 5: 659–675.

Lengwiler, Martin. 2008. *Participatory Approaches in Science and Technology. Historical Origins and Current Practices in Critical Perspective*. "Science, Technology, & Human Values", Vol. 33, 2: 186–200.

Levitas, Ruth. 2000. *Discourses of Risk and Utopia*. In: Barbara Adam, Ulrich Beck, Joost van Loon. *The Risk Society and Beyond. Critical Issues for Social Theory*. London, Thousand Oaks, New Delhi: SAGE Publications, 198–210.

Levy, David J. 2002. *Hans Jonas. The Integrity of Thinking*. Columbia, London: University of Missouri Press.

Lövbrand, Eva, Roger Pielke, Jr., Silke Beck. 2011. *A Democracy Paradox in Studies of Science and Technology*. "Science, Technology, & Human Values", 36 (4): 474–496.

Luhmann, Niklas. 1991. *Risk: A Sociological Theory*. Berlin: de Gruyter.

MacIntyre, Alasdair. 1996. *Dziedzictwo cnoty. Studium z teorii moralności*. Warszawa: Wydawnictwo Naukowe PWN, trans. Adam Chmielewski.

Marres, Noortje. 2007. *The Issues Deserve More Credit: Pragmatist Contributions to the Study of Public Involvement In Controversy*. "Social Studies of Science", Vol. 37, 5: 759–780.

McLuhan, Marshall. 2001. *War and Peace in the Global Village*. Corte Madera, CA: Ginko Press.

Merton, Robert K. 2002. *Teoria socjologiczna i struktura społeczna*. Warszawa: PWN, trans. Ewa Morawska, Jerzy Wertenstein-Żuławski, second edition.

Mills, C. Wright. 2008. *Wyobraźnia socjologiczna*. Warszawa: Wydawnictwo Naukowe PWN, trans. Marta Bucholc.

Mirowski, Philip, Robert Van Horn. 2005. *The Contract Research Organization and the Commercialization of Scientific Research*. "Social Studies of Science", Vol. 34, 4: 503–548.

Mucha, Janusz. 2009. *Uspołeczniona racjonalność technologiczna. Naukowcy z AGH wobec cywilizacyjnych wyzwań i zagrożeń współczesności*. Warszawa: Wydawnictwo IFiS PAN.

Nahuis, Roel, Harro van Lente. 2008. *Where Are the Politics? Perspectives on Democracy and Technology*. "Science, Technology, & Human Values", Vol. 33, 5: 559–581.

Nowotny, Helga, Peter Scott, Michael Gibbons. 2001. *Re-Thinking Science: Knowledge and the Public in an Age of Uncertainty*. London: Polity Press, Blackwell Publishers.

Okrasiński, Krzysztof, Agnieszka Wasilewska. 2011. *Udział na niby*. „Nowy Obywatel" 3: 58–63.

Ostolski, Adam. 2009. *Krótki kurs historii ruchu ekologicznego w Polsce*. In: *Ekologia. Przewodnik Krytyki politycznej*. Warszawa: Wydawnictwo Krytyki Politycznej, 401–424.

Papadopoulos, Dimitris. 2010. *Alter-ontologies: Towards a Constituent Politics In Technoscience*. "Social Studies of Science", Vol. 41, 2: 177–201.

Passmore, John. 2002. *Enwironmentalizm*. In: Robert E. Goodin, Filip Petit (ed.). *Przewodnik po współczesnej filozofii politycznej*. Warszawa: Książka i Wiedza, trans. Cezary Cieśliński, Marcin Poręba, 606–627.

Perelman, Michael. 2005. *Manufacturing Discontent. The Trap of Individualism in Corporate Society*. London, Ann Arbor, MI: Pluto Press.

Postman, Neil. 2001. *W stronę XVIII stulecia. Jak przeszłość może udoskonalić naszą przyszłość*. Warszawa: PIW, trans. Rafał Frąc.

Rowland, Nicholas J. 2005. *Science and Technology Studies Saves Planet Earth via Latour*. "Social Studies of Science", Vol. 35, 6: 951–954.

Shapin, Steven. 2000. *Rewolucja naukowa*. Warszawa: Prószyński i S-ka, trans. Stefan Amsterdamski.

Sismondo, Sergio. 2010. *An Introduction to Science and Technology Studies*. Malden, MA, Oxford: Wiley-Blackwell, second edition.

Slaughter, Sheila, Gary Rhoades. 2002. *The Emergence of a Competitiveness Research and Development Policy Coalition and the Commercialization of Academic Science and Technology (1966)*. In: Philip Mirowski, Esther-Mirjam Sent (ed.). *Science Bought and Sold. Essays in the Economics of Science*. Chicago, London: Chicago University Press, 69–108.

Sombart, Werner. 2001. *Ujarzmienie techniki*. In: Erhard Schütz (ed.). *Kultura techniki. Studia i szkice*. Poznań: Wydawnictwo Poznańskie, trans. Izabela, Sven Sellmer, 316–330.

Stankiewicz, Piotr. 2004. *Nauki społeczne wobec zagrożeń cywilizacyjnych*. „Studia Socjologiczne" 3: 149–168.

Stankiewicz, Piotr. 2007. *Konflikty technologiczne w społeczeństwie ryzyka. Przykład sporu o budowę masztu telefonii komórkowej*. „Studia Socjologiczne" 4: 87–116.

Stankiewicz, Piotr. 2010. *Rozdwojona tożsamość ekspertów. Między lobbingiem a nauką*. In: Bożena Płonka-Syroka (ed.). *My i wy. Spory o charakter racjonalności nauki*. Warszawa: DiG, 197–219.

Stankiewicz, Piotr. 2010a. *Co zrobić ze społeczeństwem? O władzy, polityce i (bio)technologii*. „Obywatel" 3 (50): 31–37.

Stern, Nicholas. 2009. *Ekonomia zmian klimatycznych – podsumowanie raportu*. In: *Ekologia. Przewodnik Krytyki politycznej*. Warszawa: Wydawnictwo Krytyki Politycznej, 294–348.

Stiglitz, Joseph. E. 2002. *Globalization and Its Discontents*. New York, London: W.W. Norton & Company.

Strydom, Piet. 2002. *Risk, Environment and Society. Ongoing Debates, Current Issues and Future Prospects*. Buckingham, Philadelphia: Open University Press.

Sunstein, Cass R. 2002. *Risk and Reason: Safety, Law and the Environment*. Cambridge: Cambridge University Press.

Sunstein, Cass R. 2005. *Laws of Fear: Beyond the Precautionary Principle*. Cambridge: Cambridge University Press.

Sutowski, Michał. 2009. *Zielona energia – krótki kurs*. In: *Ekologia. Przewodnik Krytyki politycznej*. Warszawa: Wydawnictwo Krytyki Politycznej, 256–276.

Szahaj, Andrzej. 2007. *Zwrot antypozytywistyczny dopełniony*. „Teksty Drugie" 1–2: 157–163.

Turner, Stephen. 2001. *What is the Problem with Experts?* "Social Studies of Science", Vol. 31, 1: 123–149.

Wallerstein, Immanuel. 2004. *Koniec świata jaki znamy*. Warszawa: Wydawnictwo Naukowe Scholar, trans. Michał Bilewicz, Adam W. Jelonek, Krzysztof Tyszka.

Wallerstein, Immanuel. 1998. *Utopistics Or Historical Choices of the Twenty-First Century*. New York: The New Press.

Welsh, Rick. 2006. *Precaution as an Approach to Technology Development. The Case of Transgenic Crops*. "Science, Technology, & Human Values", Vol. 31, 2: 153–172.

Wynne, Brian. 1992. *Uncertainty and Environmental Learning*. "Global Environmental Change" 2: 111–127.

Wynne, Brian. 1995. *Public Understanding of Science*. In: Sheila Jasanoff, Gerald E. Markle, James C. Petersen, Trevor Pinch (ed.). *Handbook of Science and Technology Studies*. London, New Delhi: Sage Publications, 361–388.

Yearley, Steven. 2005. *Making Sense of Science. Understanding the Social Study of Science*. London, Thousand Oaks, New Delhi: Sage Publications.

Zacher, Lech W. 1986. *Współczesne zachodnie rozważania o technice*. In: Lech W. Zacher (ed.). *Filozofowie o technice. Interpretacje dawne i współczesne*. Warszawa: Krajowa Agencja Wydawnicza RSW „Prasa-Książka-Ruch", 189–215.

Zacher, Lech W. 1994. *Socjologia ryzyka. Próba nowej subdyscypliny*. In: Lech W. Zacher, Andrzej Kiepas (ed.). 1994. *Społeczeństwo a ryzyko. Multidyscyplinarne studia o człowieku i społeczeństwie w sytuacji niepewności zagrożenia*. Warszawa – Katowice: Fundacja Edukacyjna TRANSFORMACJE, 20–42.

Zacher, Lech W. 2000. *Ryzyko społeczne*. In: *Encyklopedia Socjologii*, vol. 3. Warszawa: Oficyna Naukowa, 357–361.

Zacher, Lech W. 2002. *Technika a społeczeństwo*. In: *Encyklopedia Socjologii*, vol. 4. Warszawa: Oficyna Naukowa, 172–176.

Ziman, John. 2000. *Real Science*. Cambridge: Cambridge University Press.

Zybertowicz, Andrzej. 1999. *The Success of the Natural Science Sociologically Explained*. „Studia Metodologiczne", vol. 29: 11–34.

Zybertowicz, Andrzej. 2003. *„W przyszłość wkraczamy tyłem". Uwagi o cywilizacji współczesnej*. In: Andrzej P. Kowalski, Anna Pałubicka (ed.). *Konstruktywizm w humanistyce*. Bydgoszcz: Oficyna Wydawnicza Epigram, 99–102.

Zybertowicz, Andrzej. 2003a. *O zacnych i niecnych regułach postępowania (także naukowego)*. In: *Etyka w nauce*, praca zbiorowa. Warszawa: Fundacja na Rzecz Nauki Polskiej, 64–71.

Tomasz Stępień

# Nanotechnology. Assessment and Convergence inside the Technoscience

## A. Constitution of Nano-Domain as Science and Technology

### 1. The background of technological convergence: From 'normal' to 'post-normal' science

The starting point by the analysis and characteristics of nano-domain as science and technology is the process of emerging new techno-sciences such as biotechnology, information technology and cognitive sciences and the nano-based convergence between them. In order to analyse the present development in the field of converging technologies the question concerns the definition of nano-domain itself. The background of this analysis is the process of transition from 'normal' to 'post-normal' and from 'academic' to 'post-academic' conception of science. Moreover, the process of nano-domain constitution, but also the accelerated nanotechnological advance in the last decade, integrates three major and meanwhile established theoretical conceptions and frameworks of the present philosophy of science and technology: the technology assessment (TA), the concept of technoscience and the reciprocal relationships between science, technology and society (STS), and the theoretical framework of converging technologies (CT). In this manner, by answering the question – what nano is – the process of emergence of the techno-sciences in the field of natural and engineering sciences from the one hand, and the established philosophical and societal approaches from the other hand, shall be included. Following to these we have a kind of "huge perplexity" resulting from various interdisciplinary overlapping between the different disciplines and theoretical approaches, which are at the same time characterized by a prospective orientation. The explanation of nano-issues itself appears hereby as crucial for the natural and engineering sciences as well as for the humanities and social sciences, but also for science and technology at all. These "huge perplexity" of "emerging techno-sciences" has also its consequences with respect to the science understanding and the practice of research: "All this is very confusing to the researchers participating in the enterprise because their training, equipping them to solve puzzles with very strict boundaries, has left them unprepared for this new sort of science" (Kjolberg and Wickson 2010: Foreword).

Therewith one of the main tasks is to determine these perplexities and confusions emerging with the new techno-sciences by assuming which priorities there are and which issues shall be investigated, and on which problems the analysis shall be focused. Confronted with the new techno-sciences the theoretical reflection includes a set of assumptions and statements resulting from nanotechnological development. First of all it is the fact that external forces shape the framework of knowledge, research objectives, and scientific practice from the one hand, from the other hand there appears the significance of hazards of the work concerning nano-based materials, products and treatments through its whole life cycle and its impacts on humans and the environment. Following to this the analysis focuses on the guarantee of safety in the field of new technologies which are at the same time outlined by the confrontation between the public opinion and different pressure-groups. The category of safety can be identified hereby with the degree of societal acceptance of science, research and technology at all.

*1.1. Theoretical framework of nanotechnology*

From the point of view of the theoretical conception of science there is assumed that the emerging techno-sciences in form of converging technologies (nano-bio-info-cogno) are expression of the 'post-normal' science as a kind of reflective attitude and practice towards the present complexity of science and research at all. These relevance and significance of the post-normal conception of science appear in the case where "facts are uncertain, values in dispute, stakes high and decisions urgent"; whereas the traditional understanding of applied sciences is characterized above all by low degree of uncertainties and decision stakes, "where safe routine is effective and appropriate"; but the transition from 'normal' to 'post-normal' science, from low to high degree of uncertainties consists of the extension of "the peer community who assess problems and solutions", comparing to the formal expertise: "Bringing a post-normal perspective to techno-science will require radical changes in the mindset of scientists and of those who sponsor or direct their work" (Kjolberg and Wickson 2010: vi-vii; Funtowicz and Ravetz 1994: 1881). In this context are also situated the new tasks of research policy, R&D programs and research initiatives to regulate or to framework the objectives of the post-normal science. Hereby the major challenge is to elaborate a regulative framework of control confronted with unpredictability appearing at the nanoscale as well as by labs' research and mass dissemination of nano-based products. This regulative framework of control shall be focused on safe use and impacts on health and the environment. Therefore with the emerging techno-sciences a new strategic governance of interventions is needed.

## A. Constitution of Nano-Domain as Science and Technology 79

Accordingly to this there remain – concerning the technology assessment – the dilemma between 'too late' and 'too soon' scenarios, and the models 'ethics at first' or 'at last' approaches. Confronted with these dilemma and crisis involved in the present development of science and technology the conception of post-normal science, besides the 'post-academic' and 'mode 2' conceptions, appears as an alternative approach in context of the technological advance in the 20$^{th}$ century characterized by nuclear weapons, communication technologies or genetic engineering, and the controversies but also side effects involved in. Following to these experiences are established research initiatives integrating representatives of different social groups and communities focused on analyses and assessment of the technological advance with regard to the societal and environmental impacts. The background of these various initiatives among other was the constitution of nano-domain as a new field of science and research. Most of the developed approaches of a complex analysis of the nanotechnological development are focused on four major categories or nodes from the point of view of social sciences and humanities: 1) the public perception of nano-domain, nano-technology and nano-based research, 2) the governance and the political process of decision-making in the field of science and research, 3) the philosophy with the appearing general questions concerning the human being, 4) the conception of science taking into account the influence of the new techno-sciences on knowledge and scientific practice.

The nano-domain can be understood as "the places and interfaces where nano science and technology meets macro social phenomena" with the aim to explain the different perspectives and dimensions "what nano is, and what it means in a social context (…), to showcase a diverse range of social views on nano in a way that might help stimulate and facilitate dialogue between those who do nano science and technology and those who study it as a social, political and cultural phenomenon" (Kjolberg and Wickson 2010: 1). Understanding as a science nano-domain encircles a wide range of disciplines such as physics, chemistry, biology, materials and computer sciences. As a technology it concerns applications in totally different sectors and branches such as energy, transport, medicine, textiles or communication, and generally it is subsumed to the enhancement of materials properties but also to the improvement of human performance (Roco and Bainbridge 2003).

At the same time, accordingly to the statement of R. Feynman and his famous speech *There's Plenty of Room at the Bottom* from 1959 at the California Institute of Technology, the nano-domain appears as a new field of science and research, "in which little has been done, but in which an enormous amount can be done in principle (…) that it might tell us much of great interest about the strange phenomena that occur in complex situation (…) that it would have an enormous number of technical applications" resulting from the capabilities "of manipulating and controlling things on a small scale" (Feynman 1960: 22). Feynman underlined the fact of changing materials properties, so for instance "magnetic properties on a very small scale are not the same as on a large scale; there is the 'domain' problem involved", i.e. "electrical equipment won't simply be scaled down; it has to be redesigned" (Feynman 1960: 26). Following to this by entering into the nano-world "we have a lot of new things that would happen that represent completely new opportunities for design. Atoms on a small scale behave like nothing on a large scale, for they satisfy the laws of quantum mechanics. So, as we go down and fiddle around with the atoms down there, we are working with different laws, and we can expect to do different things. We can manufacture in different ways" (Feynman 1960: 36).

The nano (from ancient Greek: the dwarf) is 1 nm: $10^{-9}$ and equals a billionth of a meter, e.g. a sheet of paper is around 100.000 nm thick, and a human hair is about 80.000 nm wide, a red blood cell has a diameter of 7–8000 nm, and finally virus particles are similar in size to many nanoparticles, with maximum dimensions of 10 to 100 nm. The nanoscale is the range between 1–100 nm, but this range with its beginning and end is also questioned, because atoms and molecules are smaller than 1nm and the assumption of technological convergence presuppose the ability to operate and manipulated with molecules, atoms and nanoparticles, which "exhibit novel properties that are often vastly different from their bulk counterparts (…) the discovery of which has led to widespread interest in their potential commercial and industrial applications" (Riediker and Katalagarianakis 2010: 38). At the same time the nanoscale and nanoscience have both scientific and political (societal) relevance, "because they have the possibility to affect everything from funding, to risk assessment and product labelling" (Kjolberg and Wickson 2010: 2). Generally the analysis of nano-domain from point of view of the humanities and social sciences focuses on the phenomenon, properties and impacts associated with the 'nanoscale' and less more on a particular range of numbers. As dominant attitude is to refer to this nanometer measure unit; in consequence there are totally different approaches and projects included in nanoscience and nanotechnology. This expresses also the 'nano' as a still emerging field of science and technology, where the measure and size are discussed, because analogically which sense has a science in the size range of 1 to 100 meters for instance. But the reason "for talking about 'nano' sciences and technologies as something distinct and unique is that objects at the nanoscale may express different properties from these expressed by larger objects of the same material. Properties, such as colour, conductivity, reactivity

A. Constitution of Nano-Domain as Science and Technology        81

and melting point, can all change at the nanoscale" (e.g. the reactivity of the 'red' gold nanoparticles), and these new properties appearing at the nanoscale are the quantum effects on this scale, that "the increase in surface area to volume ratio that occurs (when an object is divided into smaller pieces, the volume stays the same but more surfaces are created, which often enhances reactivity)". From the other side "properties of larger objects are also related to nanoscale atomic configuration", so for instance graphite and diamond consisting of carbon atoms, "but these materials have very different physical properties because of the way in which atoms are arranged (in sheet form for graphite and in a tetrahedral shape for diamond)" (Kjolberg and Wickson 2010: 3–4).

One of the first areas in nano-research was devoted to discover and then to manufacture different atomic structure for carbon such as fullerenes and nanotubes, which would have high conductivity and would be stronger and lighter than steel. These are also example "of the way in which restructuring atoms at the nanoscale can create materials with novel properties"; and this point is crucial by the definition of nano-domain, i.e. "the ability to employ, engage, and/or manipulate the novel properties of the nanoscale", what means that nanoscience and nanotechnology "are seen as not just working on the nanoscale, but actively investigating and utilizing the novel properties that are in effect there" (Kjolberg and Wickson 2010: 4). At the same time the development of nanoscience and nanotechnology in the last 20 years underlines the significance of new decisive factors such as representative 'icons' and relationship to the society and public opinion. Interesting is above all the 'iconic' background of nanotechnology development: "No new development of science is exclusively limited to lab work. Rather, it must attract public attention in order to plausible present itself as important, promising and, above all, worth of being funded. Informing and self-advertising is part of the scientific business. (…) Where representatives of a new enterprise go public, attractive pictures, symbols and signals are needed. They serve as brands with a high recognition value. They are the 'icons' of the new trend" (Brune et al. 2006: 24).

Therewith the 'technological turn' with the new emerging techno-sciences is completed by an 'iconic turn', and as the icons and symbols of nanotechnology are identified: 1) the symbolical and pictographic metaphors in allusion to the title of Richard P. Feynman's speech *There's Plenty of Room at the Bottom* from 1959; 2) the letters IBM shaping by xenon atoms by Donald M. Eigler in 1989; and 3) the figure of small man formed by a few particular molecules. Nowadays there are specific geometrical icons expressing the nanotechnological 'architecture' for instance of nanotubes and nanoparticles. These icons or symbols have to express the essential of the new technological and scientific approaches, but also the new attitude towards them. The nanotechnological icon IBM remains symbolically as expression of the ability of implementation and commercialisation of the new technology. The speech by Feynman plays instead of this a programmatic approach which should be relevant to the technological and scientific development but also to the philosophy of science and knowledge at all.

The plenty room at the bottom, suggested Feynman, is a room "that you can decrease the size of things in a practical way" with respect to the laws of physics (Feynman 1960: 24). The objectives of this new approach are the new kind of phenomenons appearing at the small scale connected with possibilities of manipulating and controlling the 'things'. So for instance in the question: How it is possible to 'write' at the nanoscale? Feynman explained, that "[w]e can reverse the lenses of the electron microscope in order to demagnify, as well magnify" the 'things' (Feynman 1960: 23). It is also the paradox of miniaturization in the field of nanotechnology and nanoscale, i.e. miniaturization as 'demagnify' is the other face of 'magnify': "in a cube of material one two-hundredth of one inch wide – which is the barest piece of dust that can be made out by the human eye", so that, argued Feynman, "there is plenty of room at the bottom!" (Feynman 1960: 24). The aim of Feynman's approach was not only to miniaturize the 'things', but also to design totally new interfaces between natural and engineering sciences, between physics, biology and chemistry. Feynman underlined: "This fact – that enormous amounts of information can be carried in an exceedingly small space – is, of course, well known to the biologist (…) of how it could be that, in the tiniest cell, all of the information for the organization of a complex creature" (Feynman 1960: 24). In this way Feynman designed the passage from physics to biology at the small scale, that "all this information is contained in a very tiny fraction of the cell in the form of long-chain DNA molecules in which approximately 50 atoms are used for one bit of information about the cell" (Feynman 1960: 24). The condition of entering into the small scale and constitution of the interfaces between physics, biology and chemistry was accordingly to Feynman the development of better electron microscopes. This was also the condition of forthcoming of biology and the possibilities to view individual atoms and the DNA structure: "It is very easy to answer many of these fundamental biological questions; you just look at the thing! You will see the order of bases in the chain; you will see the structure of the microsome" (Feynman 1960: 24). At that time the microscope itself represented the interface between physics (nanotechnology) and biology; in consequence biology and then chemistry were mathematized, what can be regarded also as the background of constitution of nanobiotechnology. In this manner the aim of Feynman's approach was by the improvement of research instruments to achieve a synthesis of materials and objects: "Given the

## A. Constitution of Nano-Domain as Science and Technology

very small, highly active biological units, biology (…) has indicated a strategy for demagnifying computer components to molecular size in order to achieve any desired high performance with a minimum expense of material and energy", because at the nanoscale the problems of friction and lubrication are disappeared, so that a "hierarchy of self-producing machines with iterative miniaturization could realize the miniaturization of mobile working machines like the surgeon to be swallowed" (Brune et al. 2006: 26).

In this way Feynman anticipated intuitive the interfaces between nanobiotechnology, computer science and nano-medicine, which today make out the pivotal research fields in the nano-domain and technological convergence. Feynman designed in his speech the possibilities of manufacturing on a small scale in analogy to the biological writing and storage of information, and manufacturing of various substances: "Consider the possibility that we too can make a thing very small, which does what we want – that we can manufacture an object that maneuvers at that level!" (Feynman 1960: 25). The motive of writing and storage of information refers by Feynman's speech also to the interfaces with regard to the computer technology and the possible economic impacts what supposed the first methodological anticipation of miniaturizing the main-frame computer at that time in the top-down approach, and in analogy to the human brain. Finally, the speech by Feynman included the first design of new nano-based medical treatments (cf. Feynman 1960: 30).

With the reduction of the size and scale the new phenomenons appear, because the 'things' do not "simply scale down in proportion" (Feynman 1960: 34), i.e. materials are held together by the Van der Waals principle of attraction on a molecular level from the one, but from the other side the question is, what happens by rearranging atoms one by one?: "we can arrange the atoms the way we want", but "[w]hat would happen if we could arrange the atoms one by one the way we want them (…) that when we have some control of the arrangement of things on a small scale we will get an enormously greater range of possible properties that substances can have, and of different things that we can do" (Feynman 1960: 34). Feynman discovered in this way "something, in principle, that can be done; but in practice, it has not been done because we are too big" (Feynman 1960: 36). But the overcoming of this limit by nanotechnology would mean a totally new type of manufacturing, "if we go down far enough, all our devices can be mass produced so that they are absolutely perfect copies of one another", because at "the atomic level, we have new kinds of forces and new kinds of possibilities, new kinds of effects. The problems of manufacture and reproduction of materials will be quite different" (Feynman 1960: 36; today the best example hereby are the emerging additive manufacturing technologies, cf. Gibson et al. 2010).

There is a set of questions referring to Feynman's statements and relevant to the present development of nanotechnology. 1) What is the scope of area covered by nano-research, Feynman understood this area as both objects of everyday life and objects such as atoms and molecules. 2) How has to be defined the nano-domain as science and technology and which objectives have to be included: size/scale or properties/functions? The new realm of nano objects results from size/scale or from the new properties, processes or treatments? 3) Therewith, how far the unit of 'nano-meter' is the decisive factor of scientific description? The measurement appears as crucial in this field, but: "How should an art of measurement be designed or conceived that catches up with the miniaturization and at the same time is more precise by orders of magnitude as compared to the objects to be measured?" (Brune et al. 2006: 27). 4) The other questions concern the 'actors' in the field of nanotechnology, i.e. what are the driving forces: innovation or discovery of new effects, laws or properties by technical potential of labs and instruments, then the discoverer or inventor, and engineering basic or applied research: "Is it an inquiry into nature or rather extension of the art of engineering?" (Brune et al. 2006: 28). 5) And finally, how far nanotechnology and nano-science change the traditional understanding of physics, biology and chemistry?

In this set of questions the scope of nano-domain has to be understood multi-disciplinary, where all scientific disciplines and research areas but also implementation fields make synergistically their contributions. Following to this the nano-domain cannot be reduced only to the scaling effects and miniaturization of existing structures. In physics, biology and chemistry the nanotechnological development concerns for instance nanoscopic building blocks, quantum size effects, or molecular units as construction of more complex systems (molecular 'motors', 'machines', 'pumps' or 'switches'). All these examples are relevant for the future data storage and computer generation, e.g. quantum dot computer, molecular systems or optical computing which are storage capable and switchable. The other example are practical applications of nanosized magnetic systems used by development of nano-instruments such as atomic force microscopy (AFM), magnetic force microscopy (MFM), magnetic resonance force microscopy (MRFM), and then microscopies with a dispersion below the wavelength of light such as scanning-near-field optical microscopy (SNOM), scanning-near-field magneto-optical microscopy (SNMOM). But the major challenge by the development of nanoscience and nanotechnology is to elaborate or create the interfaces between the bio-system and the artificial nanoscaled building blocks with nano-tools and nano-devices with the first applications in medicine. Hereby remains open the question, how far the self-assembling of the living nature can be imitated by the nano-objects, especially by hybrid systems consisting of artificial engineered nano-building blocks with the molecular units. In this context appears the controversy related with nanotoxicology and the health risks and side-effects of nano-based medical treatments.

This is also the point where the philosophical and ethical contribution to nanotechnology is extrapolated. From philosophical point of view, exactly from the theory of science, it is a challenge to overcome the existing hitherto boundaries between philosophy and natural and engineering sciences with the postulate, that the "prospective thinking of scientists and the reflective thinking of philosophers might support each other in order to prevent the field from getting on a wrong track or to encourage or to promote novel ideas"; and the philosophical contribution to nanotechnology shall be critical "in a sophisticated philosophical sense of promoting sharp conceptual distinctions or well-founded judgments" (Brune et al. 2006: 29). The philosophical questions concern the quantitative determination of nano size and scale, and the qualitative one focuses on properties and functions: How far these size/scale and properties/functions are the limits of the nano-domain? In the research and laboratory practice this concerns the methodological and theoretical framework of measurement. One of the most important issues in this context is the significance of the technical laboratories' infrastructure which directly determines both the research and possibilities of applications. In this manner accordingly to the theory of technoscience the non-human elements can decisively determine and shape the theory and applications emerging in nanoscience and nanotechnology, so that there appears a new kind of relation between the 'concept' and the 'theory' confronted with the methods and applications. "Are there proper nano laws? Should nanotechnology be considered naturalistically or culturalistically, i.e. as a new image of nature or just as a new form of technical knowhow?" (Brune et al. 2006: 30). With the question about proper nano laws is connected the question, how far nanoscience and nanotechnology have epistemological and theoretical consequences for science and knowledge at all. The ethical implications are focused on the unforeseeable impacts and side effects by mass dissemination of nano-based materials, products and treatments, and on the responsibility as the major category by a complex analysis and assessment. The political implications underline the necessity to develop a legislative work attempts and legal framework of nano-research, nano-application and nano-development in the future. The question hereby is how to moderate the possible and admissible process of democratization of decision-making process in the nano-domain.

Following to the presented characteristics the nano-domain appears as a product of non-linguistic and linguistic human action with the problem of measurement and the definition of nano-size. Nanoscience and nanotechnology is a field of culture and human activity such as manufacturing but also the talk: "Even if some nano scientists enjoy the idea that they were immediately dealing with Nature herself, at least nanoscience and nanotechnology form part of cultural history, not of nature. Briefly, nanoscience and nanotechnology require reflection in terms of culture" (Brune et al. 2006: 31). Literally, translated from ancient Greek, 'nano-techno-logy' means theory (logos) of the artificially manufacturing (techne) of tiny things (nano: dwarf); but if this means a new thinking about nature? "That is to say that even the very name of the new field oppositely refers to the artificial character of the objects in question. Hence, even the inclination to think of nature or of natural 'nano objects' in the case of nanotechnology should be considered suspicious" (Brune et al. 2006: 31). This concerns for instance the 'nano-meter', the millionth part of a millimetre accordingly to the legal definition of 'meter'.

This problem of measurement appears by chemically identified molecules, proteins and viruses. The nano-meter as a unit of measurement results from the possibility and advance of technical reproducibility, and this determines the definition of nanotechnology which is not focused on size, scale and measurement, but on unknown hitherto effects, phenomenons, properties, and functions appearing at the nanoscale. But from the epistemic point of view it is crucial to know, by which element a measurement in nanotechnology can be defined, and this as a kind of philosophical or meta-theoretical reflection. In the empirical and scientific way 'measurement' can be defined by "determining a rational number indicating the ratio between a given unit and the object measured", which bases on the definition of H. Helmholtz (*Zählen und Messen erkenntnistheoretisch betrachtet*, 1877) and the postulated by him "system of operations that he paralleled to the construction of the system of numbers in arithmetic", and this "plain, intuitive understanding of how mathematics comes in into the management of the bodily world through the use of instruments became dominant in philosophy of science" (Brune et al. 2006: 32), and in the Vienna Circle by R. Carnap and C.G. Hempel is translated into the formal logic. But this empiricists approach involved a principal mistake concerning the measurement itself:

"Following the assumption that only experience can decide whether weight is a concept that proves symmetric and transitive with regard to equality, a corresponding measuring tool is needed. (…) The empiricist view of measurement simply ignores the fact that measuring tools (like tools and laboratory equipment in general) are not natural objects but artificially, 'technically' produced ones. It is their respective purpose that defines their function. And it is up to the competence of the individual scientist to know both that purpose and how it can be realized technically" (Brune et al. 2006: 33–34).

The size of nano objects consists of means of quantitative concept and measurement, and of complex system of technical purposes, i.e. the system of explicit norms technically realized, normally as the subject of protophysics. In the case of nanotechnology remains open the question concerning the progressive miniaturization controlled through measurement. From the other side "there is the question whether technical realization in nano size depends on concepts that lose their meaning in the process of reducing the size of tools or artificial objects", and the "prefix 'nano' simply indicates a mathematical proportion" (Brune et al. 2006: 34). The formulation of the laws of nature can be given through the measurement but also through proportions, so for instance in microbiology where the reproduction of organisms differs from them of measuring units. In the case of nanotechnology it is the question of validity of the legal definition based on measurement, because: "Inevitably, any tool for volume measurement meets a natural limit of tool precision (…) those extremely small concentrations lose operational meaning. Any attempt to realize such an extreme ratio would find its limits at the atomic structure of the measuring vessel"; therewith the question is how far the nano-meter has any operational meaning: "What the field is dealing with is not natural laws or natural objects or phenomena, but rather artificially produced situations that depend on technical purposes. This provides perspectives for answers to the question in how far our everyday measuring procedures in the laboratory also can be used on nano scales" (Brune et al. 2006: 35). Nevertheless till now is adopted a characteristics of nanotechnology as a field of science and research "focused on the creation of functional materials, devices, and systems through the control of matter on the nanometer scale, and the exploitation of novel phenomena and properties at that length scale" (Mnyusiwalla et al. 2003: R9).

*1.2. Methodological aspects of nanotechnology*

Nowadays the perception of nanotechnology is focused on conceptions of innovation and progress, what implies the epistemological dimension basing on the principle of methodical order[35]. From philosophical and epistemological point

---

35 The following in this paragraph considerations are situated in the TA-approach represented by TA-development in Germany, especially with the Office of Technology Assessment at the German Parliament (Technikfolgenabschätzung Büro, TAB) and the Institute for Technology Assessment and Systems Analysis (ITAS) at the Karlsruhe Institute of Technology. This approach is characterized by a practical (pragmatic), regulation-focused and political orientation (nano-governance, cf. Gammel et al. 2009), and results also from the German philosophy of technology with the 'normative' turn in the earlier 1970s (cf. Grunwald 2002). In the case of nanotechnology it seems to

of view as pivotal appears the distinction between 'discovery' and 'invention' and in the case of nanotechnology the new development is connected above all with the term of innovation. First of all we can distinguish between two types of technical progress: 1) the constructive progress where technical innovation results from methodical order, and where a kind of progressive inventions' accumulation takes place; 2) the empirical progress with totally new data opening new dimension of knowledge. The constructive and the empirical model of progress base "on the re-interpretation of known means for new ends", what allows to differentiate philosophically "between new developments in nanotechnology that are pre-planned (or at least could have been brought about by following a plan) and ones that cannot be planned in advance" (Brune et al. 2006: 38). In this sense the development (constructive and empirical progress) can be analysed only ex post, but it can be defined as "a chain of steps that consists of human actions with the rationality of means and ends as the leading principle" (Brune et al. 2006: 39), and in this manner is an expression of development's coherence.

The principle of methodical order expresses above all "the rationality of technical chains of productive actions", and in this way it concerns any technical products. But this principle differs in the case of theory building, where the initial statements of a theory have to be "rephrased in equivalent ways"; in this manner the principle of methodical order implies the question, what a theory is with respect to the formal syntactical structures of the methodical order or sense, but the decisive criterion is the "technical availability of measuring procedures" as determinants of the sequence of steps, "which can be added later on as mere linguistic definitions of further parameters". The principle of methodical order "becomes important in science where theories undergo syntactic permutations of both axioms and definitions, of propositions and instructions. Violations of this principle are consequential for the validity of theorems and even of entire theories since the order of actions in controlling theories through laboratory research are ignored", and what partially happens in the case of nanotechnology (Brune et al. 2006: 39–41).

The above mentioned methodological principles seem to be pivotal by constitution of nano-domain as science and research, where the dilemma between discoveries 'just for fun' (R. Feynman) and the principle of methodical order

---

be one of the few methodologically well-founded approaches of assessment and makes out the framework of systemic risk analyses. At the same time it would be necessary to confront this methodological framework of technology assessment with the principles of STS, especially by the practice of labs' studies and labs' ethnography, which took place in the nano labs in US and UK (cf. Kjolberg and Wickson 2010: 55–84).

as the framework of research on nanotechnology with respect to the distinctions between discoveries, inventions and applications and the role of purposes is extrapolated. The invention can be understood therewith in the perspective of applications of purpose-independent knowledge about nature, but also as the invention in basic research with new laboratory instruments (nano-microscope), which open discoveries for instance in the molecular biology. The nanotechnological example hereby is the carbon nanotube, at first produced 'just for fun' without any definite purpose and as a discovery without purpose and goal it could not be used or re-interpreted as a means. At the same time it is an expression of the endeavour of nature and a kind of discovery. Following to this there appears a tension between discovery and invention: "Scientists just do not invent anything new; they only discover how nature works (…). Hence, no inventor may legitimately be requested to morally or legally justify anything. The scientist is a discoverer, not an inventor" (Brune et al. 2006: 44). – But this is changed in the case of nanotechnology. Research works in nanotechnology take place in laboratories and depend on the use of instruments (measuring tools, infrastructure, apparatuses) form the one hand, and the nano-objects are not in the same way there as natural objects from the other hand, i.e. "they do not refer to natural objects or events at all": "Hence, 'discoveries' made in scientific laboratories always discover possible technical procedures" and are expression of human action, of scientists' in action, that the scientist "makes discoveries by means of his or her inventions"; this expresses also the means-and-ends rationality, because "the experimentum crucis for every scientific statement depends on the successful realization of ends through the choice of the suitable means" (Brune et al. 2006: 45, 46).

With the dilemma between discovery and invention is connected the question concerning 'acting' of the nano-scientists, that only 'acting', and not 'behaviour', can encircle the principle of responsibility: "We are responsible for our actions while we cannot be held responsible for our behaviour", because only 'acting' can succeed or fail, and this is obvious and fundamental in the case of empirical research. Therewith the ends and purposes in the scientific experiment "are not metaphysical entities but simply projected states of affairs towards which our actions are directed. This also implies that things (objects) do not represent purposes or ends; only states of affairs do, which are linguistically represented in the form of propositions" (Brune et al. 2006: 47), this also by designing a plan as acting in order to act with 'planning' action of second order. This is important especially in the case of nanotechnology with the distinction between technical problems and its solution and based on technical possibilities searching for applications. Finally, the role of purpose with regard to the distinction between discovery, invention and application should consider the differences between social and natural sciences: the observer and participant perspectives are characteristic for instance in ethics and social sciences, and the distinction between performance and description dominate in the natural and engineering sciences: "It is not the description of actions that brings about the objects of nanotechnology but their performance" (Brune et al. 2006: 47).

Which is then the scientific character of nano-domain? First of all it is the trans-subjectivity, i.e. the interdependence of individual persons and points of view, and the universality characterised by validity for all research objects and results. The next step by scientific and theoretical extrapolation of nano-domain concerns the distinction between basic and applied research. The starting point by distinction between nanotechnological basic and applied research is the new range of phenomena appearing at the border between classical and quantum physics. This refers to Feynman's question, how far the known techniques might be miniaturized. Following to this the ability to manipulate the objects atom by atom is included in the known laws of physics, but the condition of this is "to learn to understand the special physics" (Brune et al. 2006: 49), which governs the nano-domain. Accordingly to Feynman nano-domain concerns "the problem of manipulating and controlling things on a small scale" with the aim to develop a technology "using the ultimate toolbox of nature, building nanoobjects atom by atom or molecule by molecule", so that nanotechnology "literally means any technology performed on a nanoscale that has applications in the real world" which "encompasses the production and application of physical, chemical, and biological systems at scales ranging from individual atoms or molecules to submicron dimensions, as well as the integration of the resulting nanostructures into larger systems" (Bhushan 2004: 1). The nanoscale itself consists of molecular self-assembly such as DNA which can be used as building blocks for the production of nanostructures. Nano-scaled components such as nanotubes are fabricated by using the top-down lithographic techniques and make out the micro and nano electromechanical systems (MEMS/NEMS), which can generate effects on the macroscale.

In this context the question is, if nanotechnology concerns technical or rather natural beings, and if it is domain of culture or rather nature? What means Feynman's interpretation of nanotechnology as 'the ultimate toolbox of nature'? It is not the matter of materials, because there appears the principal distinction between 'manipulation' and 'objects of manipulation', distinction between the human action and the object of this action, and finally the fundamental distinction between culture (technology) and nature: "The technical always is the artificial that depends on human ends or purposes", so that the culture can be defined as "human intervention into given circumstances", and the nature (by Feynman) is "the material from which something (atoms and molecules) is built" (Brune et al. 2006: 49). Feynman represents a naturalistic point of view where the analogy to natural life processes is underlined, so that "it is not an entirely new field", because there are lot of natural objects and processes which can be used as a matrix by "imitating and producing nanodevices and nanomaterials", so for instance the flagella motor as a "biological molecular machine", and "when nanostructures are smaller than a fundamental physical length scale, conventional theory may no longer apply, and new phenomena may emerge", what means that molecular mechanics "is used to stimulate the behaviour of a nano-object" (Bhushan 2004: 5).

This naturalistic view of nanotechnology supposes the atomic arrangement as nature given, where physics and its new (nano) technology consists of natural laws, that are providing by nature. But this naturalistic interpretation of nanotechnology has a set of consequences: "Physics is understood as something attached to micro or nano objects independently of humans", and "stands for natural properties of natural objects", what was formulated by Heinrich Hertz (*Die Prinzipien der Mechanik*, 1894) with the postulate to constitute a philosophy of physics "according to which causal relations between natural objects have to be mapped ontological relations between mental pictures" (Brune et al. 2006: 49–50). The structure of these mental pictures is understood by Hertz as 'models', by the way Hertz introduced the term 'model' into science, and it functions today as 'theories' in the meaning of a system of propositions (by Hertz it is the representation of a theory; *Darstellung der Theorie*). This mapping of mental pictures is the objective of the present conception of converging technologies with cognitive sciences as its integral part.

At the same time the primacy belongs to the theoretical framework, because: "Only after scientists have found a 'theory' of a certain rang of phenomena engineers can start to apply theories by inventing, constructing, and building new revolutionary machines. Natural laws thus provide the ground, the fundamentum, for technical application" and this in opposition to the meaning of application as "reinterpretation of (technical) means for a new, hitherto unforeseen purpose". In argumentation by Feynman this unforeseen purpose can be interpreted as the physical knowledge as such, which remains neutral towards all the possible applications. In this sense the "naturalistic view of physics, and of science in general, ignores the technical character of any scientific research" (Brune et al. 2006: 50). Nevertheless as crucial appears herewith the significance of human practice and acting. In the case of nanotechnology it is for instance the methodological choice between the top-down or bottom-up approaches. The top-down is "the way down from macro and micro objects through miniaturization to the molecular level (and in some sense from physics down to chemistry) and a strategic handling of atoms, molecules, and nano objects considered as natural entities" (Brune et al. 2006: 50; also as an in-between of classical and quantum physics). In opposition the bottom-up approach is the so called 'argument of the thick fingers' (Smalley 2001: 76) which ignores the principle of methodical order, so that the dilemma is the choice between physics and chemistry, and in methodical sense, "chemistry provides the foundation of atomic physics, not vice versa" what is crucial in context of nanotechnology: "The supposed access to atoms or molecules by physical methods remains dependent on their definition in terms of traditional chemical laboratory techniques. So if properties of substances such as color, melting point, electric conductivity, and others change as they are reduced to nano size, any question about what substances are treated makes sense only relative to the basic chemical definitions" (Brune et al. 2006: 51). Therefore the bottom-up approach expresses more the 'culture' and practice of science, but also the historical dimension of development and learning processes constituting science and research at all.

In the philosophy of science is constructed a kind of proto-chemistry as the theory of methods' and theories' formation in chemistry, which could be able to guarantee the validity of research, i.e. it is a kind of 'theory' of chemistry as a science. Proto-chemistry but also proto-physics represent therefore a kind of philosophical meta-dimension, a theoretical basis and interpretation of chemistry and physics in the case of nanotechnology: "Also, nano objects exist for the nanotechnology scientist only because of theoretical contexts. They cannot be accessed as something 'given' without the respective framework of hypothetical theories, basic methodological decisions and purposes that function as the domain of phenomena that is to be modeled" (Brune et al. 2006: 53–54). Example of this is the designation of the objective in the nano-microscopy, i.e. are there pictures or physical objects as 'artefacts'? Feynman suggested "that microscope technology be refined up to the point where atoms can be made visible" in analogy to the everyday experience, what is visible can be described, that the optical and technical approach to an object makes it possible to define the criteria of mapping, "that the instrument is approaching the object" (Brune et al. 2006: 54). But this everyday experience cannot be used in the case of nanotechnology, because the instruments used there (scanning force microscopy, atomic force microscopy or scanning tunnelling microscopy) are "methods of producing artefacts for which not only the physical principles of data production are different but also the types of effects between a sensor and the target", and following to this the problems of the microscopy "always run the risk of using false models of direct sense perception" (Brune et al. 2006: 55), so that 'triangle' and 'nano-triangle' can be totally different, and the 'nano-triangle' remains only as an 'idea' and at the same time an 'artefact'. The instruments appear hereby as the decisive factor by building the nano artefacts, but also of shaping mental pictures about nano. Therefore the understanding of artefacts by scientists and philosophers (e.g. in the philosophy of technology) can be totally different, by the scientists the artefact is a disturbance, an obstacle to view the 'natural', the 'real' object:

> "The microscopic 'pictures' are artefacts not only in the obvious philosophical sense of being purposefully produced by humans, but also in the more sophisticated sense that they are telling us more about the tool of observation than about the objects that are observed. (…) When models are confused with what is being modeled, pragmatic coherence gets lost. It is indispensable for nanotechnology, therefore, to introduce sharp distinctions between that which is invested and what is considered the outcome" (Brune et al. 2006: 57).

Accordingly to the presented description of nanotechnology following main aspects from philosophical and societal points of view can be distinguished: 1) epistemic dimension with the 'nano' as new field of knowledge and research; 2) ethical dimension concerning the acceptance or rejection of research with consequences and unforeseen impacts involved in, what underline the responsibility and side effects problems; 3) political dimension with the question concerning the destination of strategic aims and purposes which shall serve society and technological development with the legitimation problem of the aims, ends and purposes; 4) methodological dimension with the determination of subjects and used methods including the theoretical framework and the laboratory practice, the new methods of measurement, the distinction between discovery, invention and application, and finally with the distinction between nature and technology, and between top-down and bottom-up approaches.

Following to this is elaborated a recursive definition of nanotechnology: "Nanotechnology comprises the emerging applications of Nanoscience. Nanoscience is dealing with functional systems either based on the use of sub-units with specific size-dependent properties or of individual or combined functionalized subunits" (Brune et al. 2006: 11, 62). This definition focuses above all on the technological function. But which are specific purposes in the case of nano-definition? First of all there are the cognitive purposes, i.e. delimitation or separation of nanotechnology from the existing fields of science, research and technology including "the availability of new technological capabilities, knowledge and skills" (Schmid et al. 2003: 22). Secondly the nano-definition shall extrapolate the interfaces to the other fields of science and technology, especially between physics, biology and chemistry with respect to the various applications' paths of the nano-domain. Finally, perhaps the most important definition purpose in the case of nanotechnology, the definition shall encircle the ethical, societal and political aspects what can enable to identify the research community focused on nano-domain (cf. Schmid et al. 2003: 22–23). The nano-definition refers to the size, methods and properties, and as crucial appears the reference to the methods, i.e. 'top-down' as a method of miniaturization to the nano-size, and then 'bottom-up' with supposition of the ability of direct manipulation and operation with atoms and molecules with the aim to construct new devices. And by referring to properties the definition points out the interfaces and interrelations between physics, biology and chemistry.

At the same time, accordingly to the present state of knowledge and the scope of applications, Khushf postulates instead of a clear definition the elaboration of a description or characteristics of nano-domain basing on the distinction of the three main paths of nano-understanding: 1) the visions driven, 2) the size and applications driven, 3) the political programs and projects driven: "Perhaps this is a time for characterization, rather than definition. Characterization can provide content and coherence and can define the scope and range of issues, but not an identification of necessary and sufficient conditions. (…) Thus, characterization is both descriptive and constructive, capturing where nanotechnology is now and where it should be go" (Khushf 2004: 35). In this context a complex nanotechnology assessment can be regarded as nanotechnology characterization. – Concluding, it is the matter instead of a definition to elaborate complex characteristics of the nano-domain as situated or focused on the 1 to 100 nm scale. The fact is that at the nanoscale the natural sciences are converging and manifold interfaces between materials science, physics, chemistry and biology are established. At the same time with the "molecular turn" the existing hitherto hierarchy of physics as the base, then chemistry and finally biology is replaced by a symmetrical order, the limits between the natural sciences are blurred, and in analogy between the natural and the artificial: "The metaphors of hierarchy and reduction have changed to a metaphor of bridging" (Khushf 2004: 38).

### 1.3. Dimensions of nanotechnology and the necessity of modified theory of science

The problems with definition of nano-domain are interrelated with the ongoing systemic changes in the conception of science at all. Silvio O. Funtowicz and Jerome R. Ravetz have presented the conception of post-normal science as a theoretical in-between in the theory of science referring to Th. Kuhn's paradigm change from the one hand, and indicating to the present emerging concept

## A. Constitution of Nano-Domain as Science and Technology 95

of technological (and scientific) convergence from the other hand. In addition the theoretical framework of technological convergence makes out the epistemological constitution of a socially robust and socially responsible conception of science and technology at all with the aim to initiate the process of science's democratization (Funtowicz and Ravetz 1993: 739).

The issue-driven post-normal science appears as a necessity in the case where system's uncertainties and decision stakes are high. The process of science's democratization shall be started in the cases "with ethical complexities (as in biomedical science) (…) including all stakeholders in the dialogue, for evaluating quality of scientific information for the policy process" (Funtowicz and Ravetz 1994: 1881). Following to this the post-normal science is in opposition to the traditional one: "Whereas science was previously understood as steadily advancing in the certainty of our knowledge and control of the natural world, now science is seen as coping with many uncertainties in policy issues of risk and the environment" (Funtowicz and Ravetz 1993: 739). The post-normal science is characterised by: 1) an affirmative integration of uncertainties into the system and management, 2) instead of values presupposition making them explicit, 3) instead of formalized deduction as model of scientific argumentation introduction of interactive dialogue, 4) taking into account the circumstances of science and research making, because the "paradigmatic science is no longer one in which location (in place and time) and process are irrelevant to explanations", and following to this: "The historical dimension, including reflection on humanity's past and future, is becoming an integral part of a scientific characterization of Nature" (Funtowicz and Ravetz 1993: 740).

In the concept of post-normal science is underlined the interdependence characterizing the science-society relationship, that different circumstances suppose different types of relationship above all with regard to the political process of decision-making. This is also the basic argument by the theory of post-normal science with distinction of three types of science-society relationship derived from two categories, from the level of uncertainty and then from the decision stake. The post-normal science is focused on the significance of scientific information including the uncertainty in knowledge ("that good quality of information depends on good management of its uncertainties"), and on the problem-solving strategies, where uncertainties in knowledge are completed by "complexities in ethics" (Funtowicz and Ravetz 1993: 740). In this way Funtowicz and Ravetz extrapolated the role of science in the political process of decision-making. At the same time they questioned the traditional 'social contract' between scientific community and society basing on the autonomy of academic community ('academic science').

The post-normal science assumes an extension of peer community by decision making such as citizens juries, focus groups or consensus conferences (Funtowicz and Ravetz 2003: 7). This means also an "extension of legitimacy to new participants in policy dialogues" with the aim to develop "of a genuine and effective democratic element in the life of science" (Funtowicz and Ravetz 1993: 741). Therewith is accentuated the change in science making and understanding in the last decades with the idea of socially robust science: "The challenges to science were largely in the realm of ideas. Now, as the powers of science have given rise to threats to very survival of humanity, the response will be in the social practice of science as much as in its intellectual structures" (Funtowicz and Ravetz 1993: 742). This new conception of science increases the value of social factors and circumstances of science and research; therefore it contributes to the thesis that social sciences are condition of science and technology development. From the other side the intrinsic complexity of the system changes the mode of explanation but also prediction of science and society development, so that the policy concerning science and research "cannot proceed on the basis of factual predictions, but only on policy forecasts" (Funtowicz and Ravetz 1993: 742). This is also the background of increased significance of the prospective technology assessment, especially by risk management.

> "Quality assurance is as essential to science as it is to industry (…) the evaluation of quality in this new context of science cannot be restricted to products of research; it must also include process and persons, and in the last resort purposes as well. This 'p-fourth' approach to quality assurance of science necessarily involves the participation of people other than the technically qualified researchers; indeed, all the stakeholders in an issue form an 'extended peer community' for an effective problem-solving strategy for global environmental risks" (Funtowicz and Ravetz 1993: 744).

An issue with risk as the objective of post-normal science appears at the moment where "facts are uncertain, values in dispute, stakes high and decisions urgent", and where "the epistemic (knowledge) and axiological (values) aspects of scientific problems" are interacted (Funtowicz and Ravetz 1993: 744), and where certainty and value neutrality of science and research are questioned. Today it is the case of the multidimensional process of technological convergence with new emerging fields of applications. Funtowicz and Ravetz distinguish three levels of uncertainty: 1) technical with standard procedures, 2) methodological with a complexity of various data as values and reliability, and the tension between personal judgments and professional consultancy, and 3) epistemological with uncertainties at the core of the scientific problem. All the three levels concern inexactness, unreliability and ignorance: "Post-normal science has the paradoxical feature that in its problem-solving activity the traditional domination of 'hard facts' over 'soft values' has been inverted. (...) Public agreement and participation, deriving essentially from value commitments, will be decisive for the assessment of risks and the setting of policy" (Funtowicz and Ravetz 1993: 750–751). In this way 'facts' and 'values' become complementary to each other, and accordingly to this in the post-normal science are established new forms of equity and responsibility as condition of sustainable development which are consisting of "intimate connection between uncertainties in knowledge and in ethics" (Funtowicz and Ravetz 1993: 751). Moreover, the presence of uncertainty underlines the significance of dialogue and debate by problem-solving. – In both cases, by high uncertainty and decision stake, this is the relevant situation for post-normal model of science with respect to the nanoscience and nanotechnology, "that any interaction between science and policy should be exposed to a process of extended peer review"; and the uncertainty in the case of nanotechnology concerns "how to characterise, detect and measure nanoscale particles, their toxicological behavior, and the levels and routes of potential exposures, but also related to social and ethical questions around the manipulation of matter on this scale" (Kjolberg and Wickson 2010: 11).

Besides the concept of post-normal science the analysis of the science-society relationship focus on the fact of shifting to the manners of knowledge production with postulated hybrid forms which are not integrated into the distinction between academic and industrial sciences, basic and applied research, and finally between science and technology. These are expression of the changes and development towards the model of post-academic science (Ziman 2000) and/or the so called mode-2-science (Gibbons et al. 1994). In the traditional understanding of science as dominant in the second part of the 20$^{th}$ century the societal and above all ethical dimension of science are not present in public discussion. This has changed in the 1990s with the emerging significance of social relations of science and technology with the postulated social responsibility of scientists and engineers. Accordingly to Ziman this change of attitude is "symptomatic of the transformation of science into a new type of social institution. As their products become more tightly woven into the social fabric, scientists have to perform new roles in which ethical considerations can no longer be swept aside" (Ziman 1998: 1813). The existing hitherto division on academic (basic research) and industrial science (applied research) interpreted as two different cultures is recognized as inadequate by confrontation with the accelerated technological development. But in both 'cultures' of science the ethical, societal and political dimensions of science are not discussed. In the case of academic science the official ethos "systematically shuts out all such considerations" accordingly to Merton's principle of disinterestedness and objectivity: "The important point is that this 'no ethics' principle is not just obsolete module that can be uninstalled with a keystroke. It is an integral part of a complex cultural form. (…) Academic scientists have always, of course, brought ethical considerations into their scientific work. But they have had to smuggle them in from private life, from politics, from religion, or from sheer humanitarian sympathy" (Ziman 1998: 1813). In the case of industrial science there also "is no ethical term in its social algorithm" (Ziman 1998: 1814). The other phenomenon connected with the post-academic science is the dispersed responsibility resulting from the networking and time-limited mode of research and cooperation.

In this way the presented 'alternative' conception of science with regard to specificity of technological convergence make out the concept of socially robust and perhaps socially responsible science and technology at all. In the concept of mode-2-science the analysis of science-society relation is focused on "the way in which public funding agencies have shifted from institutions primarily responsible for maintaining basic research at universities, to instruments for attaining national social and economic goals" (Kjolberg and Wickson 2010: 12). By the linear model is supposed that the funding of basic research include the social benefits, and in the model of mode-2-science shall be justified whether research and projects are societally useful with regard to public funding, so that social usefulness underlines the significance of ethical dimension in the reciprocal science-society relationship.

In both conceptions of science – post-normal and post-academic / mode 2 – the distinction between basic and applied research is blurred, and science itself appears "as a tool to be directed towards social priorities, the position of a science free from social, ethical and political scrutiny, critique and curtailment, disappears", instead of this there is "a corresponding shift in the notion of what constitutes 'good' science" from the traditional meaning with the aim of science to reveal the truth, to the understanding of science as reliable knowledge "that works and can be used and applied with success" (Kjolberg and Wickson 2010: 13). With the post-academic and mode-2 conceptions of science the 'good' science is recognized as socially robust knowledge, which is accepted by society, but what shall be underlined: "This is not to suggest that truth becomes what society wants, accepts and deems valuable, but rather that social criteria for deciding the types of questions, problems and products publicly funded science should be occupied with, is given enhanced importance" (Kjolberg and Wickson 2010: 13).

In this manner making-science means thinking about its social and ethical impacts what is expression of a socially robust and responsible science. From the one hand "science, in any case, can no longer evade its responsibility", but from the other hand this "obviously does not mean science should be given a monopoly on decision-making power", instead of this the aim is to elaborate a modus vivendi for science, "that science must be forced to abandon its splendid isolation from the affairs of the community" and the whole social life (Allhoff and Lin 2009: xi-xii). Science "needs to be cultured" in the sense of Max Weber (*Wissenschaft als Beruf*, 1917), i.e. reflection on science at all instead of a radical and strict specialization of research: "We need 'reflexive' scientists: less naïve with respect to the ideological dress enveloping their research programs; but also more conscious of the fact that the science they do rests ineluctably upon a series of metaphysical decisions" but also societal and ethical dimensions (Allhoff and Lin 2009: xii).

## 2. Science and society relationship in the case of nano-domain

In the traditional meaning science and society are separated as two different domains of culture and social life. In consequence there is established a strong and in-deep going separation between science and society, above all supported by the modern concept of science. This point of view criticizes Bruno Latour when he postulated an overcoming of this conceptual separation between nature as object of science and society as object of politics, which functioned as condition of the modern science development since the end of 17th century. But now this theoretical framework of science confronted with the technological convergence creates many problems in the relationship between science and society. This separation between science and society was artificial one, because a kind of 'pure' nature or 'pure' society have never existed, so that, following to the Latour's conclusion, we have even never been modern (Latour 1993). Accordingly to this it is important to recognize by the present accelerated development of science and technology, "that science and society interact and co-create each other" and are "entangled in a mutual process of co-production" (Kjolberg and Wickson 2010: 6).

But this mean also rejection of the linear model of science which dominated in the second half of the 20th century, and which was established by Vannevar Bush from US Office of Science and Development after the Second World War and his report *Science – The Endless Frontier* (1945). The linear model of science presented by V. Bush consisted of a social contract between science and society with unwritten expectations between society, science and politics, where science and society remained the separated fields of culture and socio-political life. In consequence this social contract has designed the policy of science and research (cf. Guston 2000: 141). The linearity of this model results from unidirectional exchange between three spheres: society and politics invest into science and science gives society and politics the new knowledge, but also innovations and technologies. The condition of this exchange was a kind of specifically understood autonomy of science without any societal and political control. Moreover, the linear model of relationship between science and society presupposed two different domains of science itself: 1) the academic science basing on public funding of basic research, and 2) the industrial science with combined scientific and financial capital, and oriented on applications and products. In both cases (basic and applied sciences) is also presupposed the ability of scientific community for self-regulation. This model is questioned by the new theories of science presented in the 1990s such as the mentioned post-normal, post-academic and mode-2 science.

*2.1. Nanotechnology and the concept of a socially robust science*

The concept of a socially robust science and its framework supposes the new contract between science and society with the new form of societal expectations on science by society. Secondly the knowledge production and the system generating knowledge shall be more transparent and basing on the precautionary principle. In this context as crucial appear the communication and information manners with the public, because a socially robust science bases on public knowledge about science, research and technological development at all, but also of the spending the public funding, what underline the principles of transparency and participation of citizens in the policy-making process in form of an anticipatory governance of science and research, and therewith technological development. Moreover, the ethical dimension becomes the integral and decisive part concerning the scope of public acceptance and funding of science and technological research. This new social contract bases on the necessity to enlarge the spectrum of participants in the discussion and the political process of decision-making concerning science and research funding, that besides scientific experts also other stakeholders and laypeople are represented, and in certain circumstances their voices shall be accepted. In the post-normal and socially robust conception of science is argued in favour of "a new form of science that is more sensitive to its role in society and its ethical dimensions, more reflexive about its potential impacts and limitations, more prepared to engage in direct interactions with members of society and more open to broader notions of what constitutes quality" (Kjolberg and Wickson 2010: 13–14).

This point of view become more relevant in the 1990s with the discussion about GMO and the mad cow disease, and last but not least this tendency was enforced by the development of nanotechnology and the progressive dissemination of nano-based products into the consumer market. In consequence, the development of nanotechnology is recognized as the principal political priority, and concerning the new social contract for science it is viewed "as one of the first 'new' fields of science that is being actively advanced through a 'mode 2' style of knowledge production, being asked to develop in a socially robust way, and being recognized at an early stage as involving both high levels of uncertainty and high decision stakes" (Kjolberg and Wickson 2010: 14). This is also the background of the development of ELSA-studies conception, but also many other interdisciplinary oriented approaches, e.g. the CONTECTS-project and nanotechnology assessment[36]. The origin of the ELSA-studies reaches to the human genome project from the 1990s, where the point of view of humanities and social sciences make out an integral part of research. Besides the human genome project the second main field of ELSA-studies was the development of biotechnology, especially in the domains of agriculture and medicine, including social critics of genetic engineering.

The main task of ELSA-studies is to manage the technological development in a socially robust direction. But, for instance, in the case of biotechnology the research on ELSA was 'too late' comparing to the scientific and technological progress, because the trajectories of development of this field of science and technology including the political development strategies and the public funding programs are designed without any respect to public opinion. In the case of nanotechnology, which appeared as political and economic priority at the end of the 1990s, the ELSA-studies seem to 'get in right' time, what make it possible to integrate humanistic and social components on all levels of research, from basic to applied research and the end-products on the consumer market. The elaboration of political framework with regard to the nanotechnological development was started with two reports in US and GB (Roco and Bainbridge 2001; Royal Society 2004), which have initiated but also institutionalized the social research on these issues. In this manner the main perspectives by the analysis of nanoscience and nanotechnology from societal point of view could be distinguished.

First of all it is the process of creating this new field of science, research and technology, and that means the history of development inside the system of science and research with regard to the ongoing political processes concerning higher education and research policy. This includes the scientific point of view,

---

36  Beside ELSA or ELSI studies there are also programs focused especially on nanotechnological issues: SEIN (Social and Ethical Interactions with Nanotechnology), NELSI as nanotechnological version of ELSI, and NE$^3$LS (Nano Ethical-Environmental-Economic, Legal and Social issues) (cf. Keiper 2007: 60; Mnyusiwalla et al. 2003: R9-R10; Allhoff and Lin 2009; Kaiser et al. 2010: 199–252).

i.e. the practice of laboratory works with the role of instruments and images, then the historical dimension with analysis of the driving forces of research, the systemic point of view with the question about the transdisciplinary framework of research, and finally the pedagogical issues, i.e. how to learn / teach about emerging technologies. Summing up, it is the question, in what way nanoscience and nanotechnology is created or established as a new cross-disciplinary domain of science and research through scientific practices and social policies as the main fields of concern.

In the second perspective shall be analyzed the image and message of nanotechnology mediated by mass media creating public opinion about. The general perceptions and opinions about nanotechnology base on metaphoric and rhetoric figures, but also prejudice fear, futuristic visions and science fiction in mass media presentations. The development of nanotechnology is represented and mediated by different mediums and can be totally different understood by the representing groups of society, so that it is pivotal to recognize how nano-domain at all is presented and represented in the public sphere.

The third perspective focuses on the visions concerning nanotechnology and its relation to the humans, culture and society. It is the matter of general philosophical issues concerning for instance the personhood but also the conception of man and humanity confronted with technological advancement, what involves the analysis of relationship between human and non-humans (B. Latour), nature and technology from the one hand, and the manifold tensions and convergences between the natural and the artificial from the other hand. Therefore the ethical issues are fully integrated in this field of reflections with the confrontation of metaphysics, epistemology, and anthropology with nanotechnological development.

The fourth perspective extrapolate the political process of decision-making, but also the ability to control and monitoring the development, use and dissemination of nanotechnology and nano-based products and treatments. The analysis of nano-domain appear as a political decision-making problem requiring new approaches to governance and regulation, what includes institutions and expertise, risk and uncertainty, the ability and possibility of public participation and engagement from the one side, and from the other side the responsibility, the driving forces and values in decision-making process with regard to the social trust and regulative framework by law.

These four perspectives are interrelated to each other, and at the same time it is a framework of the process of modeling the nano-domain as science and technology itself, including: 1) laboratory practice with the new sort of instruments and the specificity of the nano-research, 2) the decisive political driving forces of research and public funding, 3) the supposed cross-disciplinary orientation in research but also in educational practice. As pivotal in this context appear the development of new scientific practices and the higher education and research policy. In the case of nanotechnology it is above all the interrelation and interdependence between social and scientific dimensions, i.e. how far the political and social aspects influenced the development of nanotechnology, and then how far the establishment of nanotechnology in laboratories is 'mediated' by social practices, beliefs and processes.

*2.2. Political background and societal impacts of nano-domain*

The decisive impulse to establish the nano-domain was the US National Nanotechnology Initiative (NNI) started in 2000. Therewith nanoscience and nanotechnology were recognized as a priority in policy of science and research, and the origin of this program was the earlier political support of research in the materials science. At the same time this initiative expresses the general political strategy applied by constitution of knowledge- and innovation-based society with the nano-domain as its significant element. Nano-domain as science and technology "is being made not only in a range of laboratories around the world, but also in socio-political instruments such as national policy initiatives, funding programmes and educational directives", and these instruments are crucial to understand the dominant position of nanotechnology in the present research policy (Kjolberg and Wickson 2010: 27).

The origin of nanotechnology development understanding as a political priority is the materials research, where the links between national policy and industry are constituted after the Second World War. From the other side there appeared a kind of side effect, because the nanotechnological initiatives created "high expectations for innovation outcomes" without "a plan how this development should be organised" (Kjolberg and Wickson 2010: 29). Nowadays an analogical situation we have with the EU flagship projects devoted to the graphene and human brain research. The historical context of nanotechnology development focuses on two topics: 1) the significance of materials science concerning both nanotechnological research topics and organizational framework of the nano-research itself, 2) the changed understanding of innovation in the science and research policy, what also designed the first programs and trajectories of nanotechnology development. Nowadays the US political initiative on nanotechnology functions as a general model and inspiration concerning scientific and innovative programmes and politics, so for instance EU Frameworks Programs in the field of research, science and higher education (EC 2004a). The background of all these programs and political approaches is the significance of materials science for organising nanotechnology itself as a new research field. The framework of US materials research is hereby exemplary for the specificity of research context in nanotechnology: 1) the connection between or transformation of the military framework and national security with national economy and the civil framework; 2) this reformulation of materials science and research can be regarded as the base of the new program of science-based economy and then knowledge-based society, where nanotechnology plays the strategic role by development and dissemination of innovative products.

In this way 'innovation' appears as the starting point of the history of nanoscience and nanotechnology, but 'innovation' that presuppose a set of links between research (innovation) and society with regard to the different 'contexts of application' that integrate the scientific practice itself in the societal dimension (Gibbons et al. 1994). The perspective of innovation assumes the exchange of knowledge and a new kind of collaboration between different research centres but also representatives of society. On this base it was possible to create the links between science, society, politics, economy and industry. In this manner the key research fields constituted the platform of "social contract for science" (Guston 2000: 5). And historically as such field functioned materials science. Consequently academic research and science become more autonomous with the aimed benefits resulting from research itself. This linear model of innovation and relationship between science and society existed after the Second World War until the 1980s.

The linear model of innovation represented unidirectional sequence of steps: the distinction between basic and applied research, the development of a technology, and finally the production and diffusion of technology and its commercialisation (Godin 2006). This linear model is modified in form of strategic science (Stokes 1997), which is placed in-between of basic and applied research. Nevertheless the conception of the social contract for science functioned as the background of the technological development in the second half of the 20$^{th}$ century, and created the framework of the present policy of science and research, but also the framework of public funding. It was crucial in this context the transition of funding system itself, i.e. the changes from military sector to civil one, so that a political (and even societal) new framework of funding policy and knowledge production in the 1990s was needed (Ziman 2000). These changes are interrelated with the new conception of the knowledge-based economy.

Following to this is developed the 'triple helix' model of innovation basing on interrelated model of interaction and communication between universities, funding agencies and industries, understood as an interplay between academic milieu, industry and politics. This model of innovation assumes that all of the involved actors not only produce knowledge and technology but also create institutional framework of the whole process of innovation (Etzkowitz and Leydesdorff 2000). The other expression of these changes of academic knowledge and science understanding and production, and at the same time constitution of modern complex system of innovation is the conception of Mode-1 and Mode-2 science: "The transition is from relative autonomy with respect to society and from a strong disciplinary structuring of the inner life of the university (Mode 1), to a more blurred (Mode 2) situation. Here, knowledge is more socially distributed and external criteria and direct value creation become important parts of what constitutes research" (Kjolberg and Wickson 2010: 33; Gibbons et al. 1994). The model of knowledge-based economy influenced also the academic system of science and research in the 1990s, the terms such as "academic capitalism" and "entrepreneurial university" (Slaugther and Leslie 1997), and finally "post-academic science", i.e. "radical, irreversible, world-wide transformation in the way that science is organized, managed and performed" (Ziman 2000: 67), have been introduced. These changes extrapolated the necessity of a new social contract for science, where the 'innovation policy' has to integrate science, society and industry and to overcome the existing hitherto separation between them.

The above mentioned model of innovation and relationship between science and society prepared in the last time the concept of reflexive system of innovation, which encloses scientific and industrial progress but also the process of social changes. In this manner the objective of the analysis, also in the case of nanotechnology, is the socio-technical system as a whole, which dominates and shapes present science, research and society. In consequence the linear model of innovation and the dichotomy between basic and applied research, and the separation of science and society is replaced by the model of a system including these different and at the same time decisive factors, that the present research is characterized by "a constant blurring and renegotiating of such dichotomies and boundaries", and in context of nano-research and nano-innovation the question is: "how to create a nano-policy that produces developments in desirable areas without taking away the variety and open-ended character of research" (Kjolberg and Wickson 2010: 34). This includes the social and historical context of basic research development in the field of nanotechnology with a few steps of development such as R. Feynman's speech, STM-microscopy, IBM-letters, futuristic visions by E. Drexler and US NNI-research program. These steps are also the 'icons' of nano-development as science and technology.

1) First of all it is the speech of Richard Feynman at Caltech 1959 with the 'program' of shaping the world atom by atom, and with totally new possibilities

to design properties and technical functions of materials. Feynman expressed in his speech the traditional point of view of the materials physics by searching at the same time for new applications. 2) In the early 1980s Heinrich Rohrer and Gerd Binnig have constructed the scanning tunneling microscope (STM), i.e. a probe microscope with atomic resolution and the postulated by Feynman possibility to more precise control at the nanoscale. STM is a symbol for advanced material culture with sophisticated instrument's and laboratories' capacities as condition of science and technology development at all. The STM produced images as expression of a "pervasive representations" and "nanopresence" for public opinion (Mody 2004: 120). 3) In the 1990s D.M. Eigler and E.K. Schweizer have 'written' the logo IBM with a few atoms by using the probe technology to manipulate individual atoms. This IBM-text indicates above all the interests of industry, and symbolically integrated the efforts of research, science and industry. 4) At the same time Eric Drexler (1986) has presented a vision of molecular engineering with a new production or manufacturing paradigm of the whole material world with molecular 'machines' and 'assemblers', self-replicating and building the material world from bottom-up. The vision presented by Drexler included both the radical positive potential as well as radical risk (auto-replication of nano-assemblers) of nanotechnology, but it has had consequences and contributed to the development of nano-research policy and formed the public opinion about nano-domain. 5) At the end of the 1990s is established the new research policy and new possibilities of co-funding the science with the first research projects on large measure such as the US National Nanotechnology Initiative (NNI) with the objective to operate and to manufacture of nanoscale objects, devices and systems.

The changed understanding of science at the end of the 20th century expressed also its openness to the public sphere. This means above all that "legitimacy for allocating resources to research has increasingly been sought using reference to civil and economic rationales. Funding of research on nanotechnology is subject to an open competition with other societal areas and goals. Hence public contexts have become increasingly important for both patrons of science funding and for researchers" (Kjolberg and Wickson 2010: 37). This concerns especially materials science, besides nano- also surface science and microelectronics, and their connection to the usefulness: "The notion of materials combines natural science and the humanities; it combines physical and chemical properties with social needs, industrial or military needs. From this coupling of natural and human aspects embedded in the definition of materials follows a basic feature of materials science: knowing and producing are never separated. Materials science couples scientific research with engineering application of the end-product"

(Bensaude-Vincent 2001: 223; quoted by Kjolberg and Wickson 2010: 39). In the case of nanotechnology the leading objective of materials science has been to find out 'the scale' that governs material properties, and this assumption functioned also as the main motive by establishing of interdisciplinary laboratories and research teams, and in end-effect contributed – above all in the US – to the establishment of the political and technological system of science and industry.

This characteristics of materials science resulted from the new research objectives and funding strategy after the end of the Cold War, i.e. besides military and space branches there appears a third one: industry and mass consumption of the new materials. This new understanding but also constitution of science and research system is started by referring to the interdisciplinary paradigm realized in form of Interdisciplinary Labs and then Materials Research Labs, principally in the fields of microwave components, semiconductors, laser technology, thin-film technology, and infrared optics. In the 1990s were added to this complex the human genome program, biotechnology and nanotechnology. Till now the aim remains hereby to bring together the "discovery-driven culture of science and the innovation-driven culture of engineering" (Parker 1997: 1). In this sense "the most exciting and important advances occur at the interfaces between traditional disciplines, forever altering the scope and boundaries of those disciplines" (National Research Council 2007: 34). – Summing up, the US National Nanotechnology Initiative represented an exemplary implementation of science as the endless frontier integrating three major elements: 1) the legitimacy of basic research in public opinion, 2) the guarantee of resources for basic research, and 3) the integration of researchers' community in both basic and applied sciences.

The development of nano-domain as a socially robust model of science integrates the political and societal contexts on the basis of an interdisciplinary orientation, what means also integration of social scientists into nano-research and nano-labs. This is also the background of the emerging significance of laboratory-based social sciences integrated into the scientific practice: 1) by the extension of the cross-disciplinary research including social sciences and focusing on the development of socially robust technologies, where the research on nanotechnology is an exemplary case of this cross-disciplinary collaboration, and then 2) by critically questioning the scope of cross-disciplinary inclusion of social sciences into the nanotechnological basic research, i.e. if it functions, which goals can be assumed with, and what potential is connected with such collaboration. With regard to the US NNI from societal point of view the main issues hereby are: 1) the open question how to ensure the responsible development of nano-domain; 2) which means should be applied to ensure the social trust and regulations by law in the case of nanotechnology; 3) how has to be denoted the scope of cross-disciplinary research integrating different stakeholders, social groups of interests, and disciplines; 4) the first area of concerns focused on anticipation and regulation of risks resulting from dissemination of the nano-engineered materials; and 5) the second area of concerns is devoted to the social and ethical impacts of nanotechnology. Accordingly to these concerns the social sciences play the mediator role between science, technology and society: "The assumption is that the social sciences will work as mediators, facilitators and representatives for social and ethical issues, helping to 'create' technologies that are more closely aligned with wider society's concerns and interests, thus avoiding the political and economic costs of public controversy", so that "the social sciences and the public become if not synonymous, then at least largely overlapping realms, and inclusion of one means inclusion of both" (Kjolberg and Wickson 2010: 56). The main steps and at the same time issues hereby are: 1) the context of science and research policy, 2) the theoretical and methodological framework, 3) the artefacts or nano-devices and the scope of their implementations.

Ad 1) Policy of science and research. The process of constitution of the nano-domain depended on the political context within the introduced principle of interdisciplinarity. Nanotechnology as a field of science, research and industry become the strategic factor in the policy-making process concerning science and research at the academic level, especially inside the policy of various R&D programs planned and started on the national, international and supranational level (so in the case of the EU Framework Programs). The first program including also the societal issues such as a responsible development was the US NNI, launched in January 2000. The development of nanotechnology understanding as a strategic political and public investment priority started in the 1990s after the Cold War period, and was enforced by the changes in the relationship between science and society, i.e. the openness of science and the new conception of socially robust science and technology besides the dominant hitherto military issues. Consequently science become public or entered into the public sphere, where the investments into R&D programs become a public issue and objective of public opinion.

The pivotal document is hereby the *US Nanotechnology Research and Development Act* from 2003 with two main assumptions applied into the US NNI, that nanotechnology shall contribute to foster "rapid technological implementation" and conduce to "technology development with more effective regard for societal consideration" (Fisher and Mahajan 2006: 5). The US NNI was the forerunner programme that has been then implemented also in Japan and South Korea in 2001, and EU in 2002 (the 6th Framework Program, cf. Roco 2003). In Europe this integration of social sciences was enforced in the 1990s by the mad cow disease with the public controversy about the genetically modified foods and the postulate of greater public participation with regard to the planned public funding and investments in science and research policy. In consequence, the principle of self-regulation of science and research is radically questioned and rejected by public opinion, and the relationship between science, society and government is renegotiated (Gibbons et al. 1994; Gibbons 1999; Guston 2000; Jasanoff 2005) from the one hand, and from the other hand there is constituted the conception of nanotechnology as a socially robust science and technology at all (Macnaghten et al. 2005). This historical context is important by explication the political and societal framework of responsible development of nanotechnology, understanding also as anticipation and prevention resulting from the experiences with technologically caused catastrophes in the last decades. This renegotiating "is against this backdrop of demands for simultaneously demonstrating more directly the social utility of research, and opening up the criteria of what would count as socially desirable technologies, that nanotechnology policy took shape" (Kjolberg and Wickson 2010: 59), and the main issue hereby is to analyse and to anticipate the impacts and risks of nanotechnology concerning environment, human condition, privacy and enhancement, but also the scope of military use and mass consumption of nano-based products (in analogy to the mentioned ELSA-study or the Human Genome Project). This motive of responsible development and societal co-creation of nanotechnology is fundamental for the EU strategy of nano-development (EC 2004a; Royal Society 2004).

Ad 2) Theoretical and methodological framework. The extrapolation of the theoretical and methodological framework from point of view of social sciences with regard to the specificity of nano-research is first of all oriented accordingly to the presuppositions and assumptions presented by science-technology-studies (STS) with underlying role of social scientists in the reality and practice in the space of laboratories. These studies take the form of lab and cultural studies of science. From political point of view it is obvious to integrate social sciences into the ongoing research in the natural and engineering sciences, and therewith to enlarge the scope of interdisciplinarity with regard to the envisaged socially robust nanotechnology. The matters are complicated by the realization of this integration of social scientists into the laboratory practice and in the technical endeavours (Barry et al. 2008). By the way, the term 'science-technology-studies' functions hereby as an umbrella of all theoretical approaches with methodologies basing on the assumption that the relationship between science and society take form of a collaboration, co-creation, coproduction and co-construction (Jasanoff et al. 2004).

The origin of social studies on science and technology were the analysis of institutions of science and research, and the structures of scientific career, but without analysis or considerations of scientific method or content of scientific knowledge. This attitude, dominant until the 1970s, has changed by emerging of the science-technology-studies with the research objectives such as the production of scientific knowledge from the one hand, and the experience with scientific knowledge in everyday life from the other hand. Following to this the so called 'laboratory studies' become one of the research methods, also understanding as an entering of social scientists into the labs. From the methodological point of view the STS-approach bases on the study of foreign tribes developed in the ethnography (Geertz 1973), and in this sense the STS-study can be derivate from the cultural turn in the second part of the 20$^{th}$ century (Bachmann-Medick 2006). The ethnographic method consists on participative observation and interviews conducted in the milieu of scientists which is characterized by own cultural norms and conventions, and in this manner is a form of tribe community which has developed the "cultural apparatus of knowledge production": "By placing science within culture, and thus within society, these studies have shown that scientific objects are not 'natural givens' but rather 'technically' manufactured in laboratories (…) and also inextricably symbolically and politically construed" (Knorr Cetina 1995: 143, 144; Knorr Cetina 1999). Following to this the scientific 'facts' are produced accordingly to the patterns of a local 'culture' as social construction (Latour and Woolgar 1979; Hacking 1983). These are also the pivotal elements of scientific process and knowledge production, which are not included in the research about science and technology hitherto.

In this manner the 'social' and the 'cultural' factors represented by milieu of scientists play strategic role by science and technology development and constitute the specificity of science 'culture' as a process of science-making (Latour 1987; Jasanoff 2004). This concerns also the ongoing research in the nano-domain, that values, norms and beliefs of a scientists' community are integral part of research itself: "These are not a sign of bias or lack of objectivity, but are instead part of the culture of different research laboratories, which are themselves part of a larger society that prioritises and funds research at the nano scale. Importantly, as the science develops and produces further knowledge the society changes" (Kjolberg and Wickson 2010: 62). In the case of nanotechnology the interconnection between science development and social change extrapolates the example of nano-based medicine, which can change not only the medical practice or treatment but also can influence the personal self-care. Therefore, the aim of the STS-study is to extrapolate the manifold interrelations between science and society by the process of becoming of a society, and by constructing natural

## A. Constitution of Nano-Domain as Science and Technology 111

and social realities. Therewith the comparative and interdisciplinary oriented lab studies, besides the analyses of 'cultural apparatus' of scientific activity, postulate at the same time an intervention into lab-practices and process of knowledge production. Following to these statements the role which humanities and social sciences can play in science and technology analysis, e.g. nanotechnology, are: 1) to be "partners with different perspectives" (Maienschein 2002: 142), 2) to be mediator by public communication and by moderating public discussion about technology, especially in the case of nanotechnology there is the concept of "trading zones" of shared expertise (Gorman et al. 2004: 69–71), or with fostering of reflexivity on the part of the scientists so as to create "governance from within" (Fisher et al. 2006: 485).

Concerning the role of social sciences in nano-research there is underlined the necessity of a "radically different" approach. In consequence social sciences shall "render scientific cultures more self-aware of their own taken-for-granted expectations, visions, and imaginations of the ultimate ends of knowledge, and these more articulated, and thus more socially accountable and resilient" (Macnaghten et al. 2005: 278). The main elements of ethnographic lab-study are: 1) the analysis of the cultural apparatuses of knowledge production including the tacit understandings, practices and processes which shape scientific practice in the nano-research and nano-labs; 2) description of implicit values, decision, beliefs and cultural expectations by the 'making of nano'; 3) the social and ethical issue in nanotechnology, i.e. the meaning of the term (Lewenstein 2004), but outside the practice of lab-works there are mentioned: the social scientists as mediators between science and society, but also as educators of the public about science, and the scientists about public opinion.

Ad 3) Artefacts or nano-devices and the scope of their implementations. This issue concerns directly the communication with the public and the significance of public opinion by nano-development. With the postulated 'opening up' of nanotechnology towards social sciences there are sustained four misconceptions and assumptions: 1) that we have "one 'public' with homogeneous and predictable views"; 2) that "the social scientists can mediate between science (or the lab) and the public"; 3) that the public "is potentially antagonistic towards the laboratory (and conversely, that being better informed will result in greater acceptance)"; and finally 4) that "science is only social to the extent that it has to accommodate or take this 'public' into account" (Kjolberg and Wickson 2010: 75). In the first period the integration of social sciences into labs based on the assumption that it by itself automatically would be lead to "openness, disclosure, and public participation" (Roco and Bainbridge 2001: iv), so that "a nanotechnology laboratory that contains social science will be more responsive to public concerns and aspirations, and thus more responsible, and less likely to generate future public resistance to commercialisation of technology"; but the experience from ethnographic lab-study underlines the opposite point of view: "It is in the framework of this concern with a resistant public that turning of the social scientists gaze towards the inside of the laboratory is seen as potentially dangerous" (Kjolberg and Wickson 2010: 76). These assumptions do not consider the power of asymmetries dictate of objectives and research agendas (cf. Joly and Kaufmann 2008) and refers to the ability that social sciences can directly influence public opinion and attitude towards technology. In general the turn to the labs-study should concern the basic research of social sciences and does not be limited to description and analysis of the laboratory practice as in the case of social sciences understood only as applied research (Barry et al. 2008). Therefore the slogan of interdisciplinarity seems to be only political one, the alternative approach are labs social studies understood as basic research in form of ethnographic study of scientific knowledge production. To the main obstacles by implementation of the principle of interdisciplinarity belong the 'power asymmetries' between social and natural sciences concerning funding but also evaluation of research and ignorance by the current research policy, and then the institutional division on natural and social sciences.

*2.3. Nanotechnology and the public opinion: Politics of nano-images*

One of the most important issues with respect to the societal dimension of nanotechnology and the technological convergence are 'images' or "the importance of imagery in the making" of nanotechnology from both sides social and natural sciences with distinction between the construction of scientific images of the nanoscale, and then the use of nano visualisations in the documents promoting and communicating the political framework of nano-research; herewith is to underline the "differences between the views and modes of analysis of social and natural scientists and develop a method for organising the work of interpreting nano images that can be used across these different views" (Kjolberg and Wickson 2010: 25). The presented hitherto nano-images play an important role by both scientific development and the public understanding of nanotechnology. But the question is what can be really 'see' at the nanoscale mediated by the nano images, and which are the messages with respect to the laboratory practice and the responsible governance in the nano-domain, i.e. how far the nano-images create the political construction of nano-domain?

Concerning the nano images two perspectives can be distinguished: 1) the image (presentation) of nanoscience at all, and 2) the image (presentation) of nano-research including ethical, legal and societal aspects (ELSA-study): "Whereas the nanoscientist may ask what object or state of affairs on the nanoscale is evidenced by the image, the ELSA researcher might be more interested in questions such as: What can I learn about the research and the researchers making this image, in terms of their objectives, world views, cultural conventions, interests or values?" There is a reciprocal enrichment between these two perspectives, "that ELSA research can be enriched by an understanding of the scientific reading of nano images" from the one hand, and the nanoscientists "may learn from ELSA research in order to arrive at ethically and politically sound use of images" from the other hand (Strand and Birkeland 2010: 85).

Nano-images are not only subject of scientific nano-research but also a motif of education and research policy using in documents, in the popular culture and mass media, as far as in the emerging nano-art. With regard to the image the question is, "if nano images really should be called images" (Strand and Birkeland 2010: 87; cf. Pitt 2005). The confusion appears by the terms 'image', 'picture', 'icon', and 'visualisation'. By nano-image it is better to use the term 'illustration' as a graphic material: "something that visually explains or decorates a text", so the term 'image' express 'representation' and/or 'interpretation' of objects or states of affair. But the question is how far an 'image' has a reference in the real world. The explanation of image as a reference makes out the base of interdisciplinary research on the point of junction between nanotechnology and its social and philosophical analysis: "The argument would be that even if one does not believe that nano images really can represent nano objects, it is still highly relevant to understand if, when and how such images are understood as representational by nanoscientists. (…) What we aim at, is an exchange of some of the knowledge and skills involved on both sides" (Strand and Birkeland 2010: 88) with the distinction between: 1) political aspects of images, 2) public interests including in various nano brochures explaining and popularising the new field of science and technology at all, and 3) how the images were constructed and used by implementations of certain motives, i.e. what is the message of a nano-image.

The political aspects of nano images refer not only to the institutions and the processes of decision-making, for instance in the field of science and research policy, but also "to processes in society involving or relating to power or authority, for instance with respect to decisions on rights and duties, wealth and welfare. In this sense, images can be made, used, interpreted or passively received in a political manner and with a political influence" (Strand and Birkeland 2010: 89). One of such examples is the EC-brochure *Nanotechnology – Innovation for tomorrow's world* (EC 2004b), originally published by the German Ministry of Education and Research. The main purpose of this brochure was even to 'illustrate' to the public opinion what nanotechnology in general is and could be in the future. In this way the brochure included images as expression of objects and states of affair at the nanoscale, images of nano labs and instruments, but also the potential applications combined with images of macroscopic objects, and finally there are also artistic imaginations about nanotechnology at all. The composition of the images' presentation appears as a collage mixing together different aspects, that nanotechnology is essentially natural but also man-made technology, and therefore as an advanced technology, which refers to all scale from nano to the astronomic scale, this is very important allusion that nanotechnology means a continuity and extension of the space technology, so for instance in the US National Science and Technology Council brochure *Nanotechnology – Shaping the World Atom by Atom* from 1999.

Consequently three types of images' juxtaposition are used by constructing of the message: 1) Nano-image and image of instruments and scientists with the message that the measuring devices makes it possible to 'see' nano, what often is not true, because the nano-images can result from processing of the raw data by software. 2) Nano image and image of plant, animal and micro-organism what expresses the aim to put in relation nano and the nature. This aspects concern above all risks and safety use of nanotechnology, because the message underlines the understanding of nanotechnology as something natural. 3) Nano-image and image of macroscale such as Earth, stars and galaxies where the nanoscale is combined with astronomic objects. Hereby as representative appears the brochure by US National Science and Technology Council (1999). How to explain this juxtaposition? Above all the nano-domain appears here as prolongation and continuation of space technology development and as a radical paradigm change, as a 'breakthrough technology': "one that break through the dam of conventional wisdom and slow progress, opening up prospects for transformative change" (Nordmann 2004: 48). In consequence nanotechnology is identified with space-travel as a symbol of the new world and new human capabilities to change this world. The background of this message is the statement of science as the endless frontier presented by V. Bush in 1945, but also the intuitive anticipation of nanotechnology by R. Feynman's speech from 1959.

The presented analyses of the nano-images concern above all the non-scientific literature. In the scientific context the nano-images should represent real objects or states of affair. But with G. Frege there is a disagreement concerning the nature of reference and representation. We can distinguish two aspects of representation: 1) the aspect of denotation (or reference) with the question, if the image is a representation of a real object, and 2) the aspect of connotation, i.e. "what claims are made in the image about the qualities and properties of the

A. Constitution of Nano-Domain as Science and Technology   115

represented object?" (Strand and Birkeland 2010: 95). The third aspect concerns the construction, i.e. the information mediated by an image as a constructed reality, and as an evidence of bias or ideological interpretation.

As example can serve the IBM-image by Donald Eigler: on the denotation-level there are 35 xenon atoms individually positioned by Eigler with the scanning tunneling microscope, but if we really can see the atoms? On the connotation-level the xenon atoms are blue and like cones, but this cannot concern the point of view of physics, i.e. 'color' does not refer to individual atoms, so that it is defined by a vision or interpretation. Following to this the interpretation by a physicist could be quite different: "The image denotes (shows) 35 xenon atoms positioned with precision in array. The image connotes that the density of the signal detected by STM tip from each atom has the shape of a cone" (Strand and Birkeland 2010: 98). Finally on the construction-level it is not the subject of physics, but this image has political aspects, i.e. the blue is the colour of the IBM as a representative of industry and is connected with the idea of the new breakthrough technology with IBM as expression of 'nano' in the future and the possibilities of marketization. This IBM nano-image can be misinterpreted by public opinion, e.g. that xenon atoms are something like blue cones, so that there appears a moral dilemma, how to 'represent' the nano image to the public, because attraction the public can result in misinforming on the connotative level: "The makers of brochures often will have to use ready-made nano images produced by scientists, so the moral dilemma should also apply to the scientists themselves" (Strand and Birkeland 2010: 99).

The analyses of the nano-images concern the two main methods of nanotechnology: scanning tunneling microscopy (STM) and atomic force microscopy (AFM). But there still remains the question if nano-images are really images of an object: "Obviously, whatever experimental method one uses at the nanoscale, images are always an assembly of signals recorded with an instrument that records some kind of interaction with the sample either than electromagnetic radiation. For some philosophers, this means that nano images are not images in the ordinary sense, and should be called as such" (Strand and Birkeland 2010: 102), because they "do not allow to us to see atoms in the same way we see trees" (Pitt 2004: 157). But in the case of nanotechnology the technology itself, the instruments, mediate the image: "We do not have to rely on technology to see the tree and check the image", if it is a good representation or not. This remains relevant for the philosophical realism, but in the other philosophical point of view the considerations are focusing on the theoretical concepts of knowledge understanding as "adequate, useful and reasonable ways to summarise evidence from experiments and experience", and following to this "the whole distinction between 'real' images and 'not really images' is meaningless. For them, 'atoms' are little more than very useful names anyway" (Strand and Birkeland 2010: 102–103). Confronted with these philosophical dilemma the scientific interpretation of a nano image "always assumes that the image only connotes a limited number of important features of what it denotes", so that "images are always a kind of shadows of the 'real' objects, at best"; and in this sense the interference between the aspects of denotation, connotation and construction underlines that "the intended connotations of nano images in terms of properties at the nanoscale are irrelevant". But by the discussion about risks and hazards, uncertainty and control of nanotechnology it is important "if one believes that it is possible to actually 'see what is going on' at the nanoscale" (Strand and Birkeland 2010: 103–104), because pictures cannot lie, but have to be read carefully.

The last and important elements by the constitution of nano-domain as a new interdisciplinary field of science and research on the point of junction between natural and social sciences is the progressive process of elaboration basic scientific literature understanding also as distinction and categorisation of the main topics with regard to the societal aspects of nanotechnology. With the elaborated basic literature it was possible to establish this new research field and to contribute to its development accordingly to the technological advance. With the emerging of nanotechnology as a new field of science and its development in the last two decades we can observe increasing number of scientific publications devoted to the societal aspects and impacts of this technology. Therewith the scientific constitution of nano-domain includes also the development of social science literature. Above all since 2005 increased the number of mutual citations by the social scientists concerning nanoscience and nanotechnology. The scope of interdisciplinarity is enlarged in the last decade by the humanities and social sciences which have posed the questions related to the societal and environmental impacts and risks resulting from development and dissemination of the nano-based products, materials, treatments and processes. Following to this nanotechnology can be applied practically into all fields of culture and social life, and interpreted as one of the main research objective in different disciplines, "where societal assessments have been embedded": "Importantly, in this round, the scope of societal assessment has been broadened, incorporating not only ethical, regulatory and legal concerns associated with nanotechnology, but also calls to embrace issues of economic impact, equity, privacy and security, environmental effects, public deliberation, public perception, and the role of media and culture" (Shapira et al. 2010: 597; Mnyusiwalla et al. 2003; Bennett and Sarewitz 2006).

In this manner the societal implications of nanotechnology constituted a significant research field enclosing science and policy with regard to the phenomenon of convergence. In this context three main positions can be distinguished: 1) the conception of converging technologies (nano-bio-info-cogno) and the US-initiative presented by Roco and Bainbridge from 2001 with the assumption, "that the disciplines engaged in nanotechnology are increasingly likely to coalesce together as the field develops" (Shapira et al. 2010: 597). 2) From the other side there are not registered remarkable increase of collaboration between disciplines integrated or associated with nanotechnology (Schummer 2004). 3) Finally it is the fact that nanotechnology with the research objective placed at the 1–100 nm scale is obviously a multidisciplinary field of research, "albeit not substantially more so than other recent emerging fields that make use of knowledge in multiple areas of science and technology" (Shapira et al. 2010: 597; Porter and Youtie 2009a,b).

Meanwhile an own body literature with theoretical and methodological framework is developed by the humanities and social sciences devoted especially to the societal phenomenon of emerging technologies, including increased interests to the nanotechnology. The main categories or dimensions of the emerged social sciences literature on nanotechnology are: 1) Trajectories of technology development and implications with assessments' and visions' development, then with issues such as convergence and societal significance. 2) Governance: with the policy making and risk assessment as the objective of social sciences and nanotechnology with focusing on philosophy, sociology, and politics. 3) Public perception and deliberation concerning nanotechnology with the problems of public opinion and attitudes, and social trust. 4) Ethics "addressing questions related to the exploitation of nanotechnology including malevolent uses, impacts on equity, abilities to control new technologies, and moral principles for decision-making" (Shapira et al. 2010: 604; Mnyusiwalla et al. 2003). 5) Science and technology studies (STS) with regard to nanotechnology, and "probing what are, or should be, the underlying matters of concern in the development of nanotechnology, including how nanotechnology objects should be described and conceptualized and how nanotechnology fits within broader projects to use science and technology for political, economic and national ends" (Shapira et al. 2010: 604). 6) Science visions as expression of fictional, futuristic or science fictional narratives of applications and dangers involved in. 7) Science mapping: patterns of interdisciplinarity, cooperation, establishment of networks and the degree of knowledge exchange and innovations, but also the role of institutions, processes and interactions confronted with technological change concerning science and technology policy (cf. Shapira et al. 2010: 603–605).

## B. Nanotechnology and Development of Assessment Regime
### 3. General issues of risk assessment by converging technologies

Nanoscience and nanotechnology underline the fact of the fundamental change in understanding of science and knowledge resulting from the technological advance in the last decades of the 20$^{th}$ century. The major challenge in this context is to deal rationally with the unforeseeable consequences and impacts resulting from scientific and technological development. Therefore the question is which methods shall be used by elaboration of recommendations as the background for political and legal acting in public sphere confronted with the emerging technologies. The significant determinants hereby are interdisciplinarity and the scope of convergence between different scientific disciplines and technologies. At the same time there appears the necessity to elaborate an assessment regime confronted with the accelerated advance of nano-, bio- and informational technologies from the societal point of view. Following to this the framework of assessment regime is recognized as a condition for societal acceptance of technology and science, but also as a condition of its development. Nanotechnology appears hereby as the strategic field in engineering sciences with respect to the new technological solutions, and is identified with innovations which concern: 1) nano-tools, 2) molecular assembly processes, 3) new conception of information and communication technologies, and 4) interfaces between natural (biological) and artificial (synthetic) systems.

Ad 1) The development of nano-tools and research apparatus is condition of the exploring and operating at the nanoscale, and with the spectroscopic tools and higher resolution electron microscopic devices was possible to construct new laboratories' practices with discovering and observations of bio-systems and physical structures, layers and phenomenons, for instance how cells work and how it is possible to operate at this scale.

Ad 2) The molecular assembly processes contribute above all to the elaboration of a new generation of materials and devices. Nanotechnology with the new laboratories' tools and devices opened the observations of the molecular systems, how they "interact in a predictable way in order to generate manifold structures and functionalities which range from simple control and molecular recognition to complex properties", so for instance "the self-assembly of three-dimensional electronic circuits", or the new concept of information technology basing on "optics and solid state electronics"; these self-assembly strategies make possible the micro- and nanofabrication with new "materials and devices with integrated nanoengineered building blocks" (Brune et al. 2006: 2).

Ad 3) Besides the new materials the major promise concerns the future nano-based information technology and nanoelectronics devices, what means also the change of paradigm by information transmission and the necessity to elaborate a new logical concept of information at all.

Ad 4) The technological convergence with the construction of interfaces between biological (natural) and engineered (synthetic) systems belongs to the crucial challenges in the nanoscaled research works. That means to discover or construct: 1) interfaces between living, natural and synthetic systems above all with regard to the biomedical applications, what is expressed by establishment of nano-biotechnology as a new research field, and this presupposes the construction of portable devices, e.g. in the field of nanotoxicology. 2) Interfaces between physics, chemistry and biology as expression of the interdisciplinary framework as constitutive principle of nanotechnology itself. 3) Interfaces between engineering sciences, biosciences and medicine, which concern the understanding of cells processes and functions by using the new nanotechnological tools and devices with the development of new possibilities of medical diagnosis and treatments, for instance by diseases detection. The condition of these constructions is the development of new nanoscale research devices for in-vivo and ex-vivo analyses.

Following to these considerations the risk assessment and risk management represent the answer to the public debate about nanotechnology focused on the side effects and unforeseen phenomena involved in technological development. On the basis of the general technological risk analysis the question concerns: 1) the specific risks and challenges involved in development and implementation of nanotechnology; 2) the properties of nanoparticles, their possibilities of transportation, proliferation and impacts on health and environment; 3) the significance of the precautionary principle in the nano-risk assessment and management; 4) the issues appearing in public debate about nanotechnology which was at the first time of nano-development focused on visionary and speculative scenarios, so for instance the metaphors of nanobots or self-replicating nano-machines, and which have determined the public perception of nanotechnology without any scientific background; 5) finally, it is the question what kind of recommendation are crucial confronted with possible risks and unforeseen side effects of nanotechnology.

Nowadays scientific progress and technological development are conditions of the economic prosperity, social development and cohesion, so for instance in the field of materials production, technologically supported medical treatments, and by enforcement of the framework concerning sustainable development. But at the same time the accelerated development and mass dissemination of new technologies accentuated the new penetration of technological risks above all in the societal dimension, which frames the public discussion and attitude towards technology. At the same time there appears the ambivalence of technological and science-based development understanding as both the condition of wealth and as bearer of risks and hazards. The driving forces of technological development are prospective oriented, i.e. the anticipation of future makes out the decision-making process on technological development, but this includes both the intended and unintended or side effects. Hereby the intended effects design the decision-making in technological development including the aims, purposes and functions, and the unintended side effects can result from misuse. Development and use of technology integrate this ambivalence.

The risk itself is a multidisciplinary object of research with two different perspectives: 1) the objective perspective as dominant in natural and engineering sciences with the probability of the side effects in form of a comparative assessment of different types of risks; 2) in the subjective perspective is underlined the normative and evaluative judgement as a base not only of risk assessment but also risk analysis as subject of public debate. Following to this distinction the risk assessment plays an important role as the subject of public debate and is involved in the societal processes interrelated with technological progress, so that 'risk' should be anticipate as a research subject of social and natural sciences, humanities and engineering sciences. Concluding, the objective perspective of risk does not integrate the individual (subjective) elements of judgement, and the subjective perspective could be interpreted as expression of private beliefs without any intersubjective dimension. The alternative distinction is between 'perceived' and 'calculated' risks, what includes both subjective and objective perspectives and aspects by risk analysis: 1) the ethical analysis of risk with normative and evaluative aspects and the question of risk acceptance, "that the normative acceptability cannot be derived from empirically measurable acceptance"; 2) the 'proximity' of the effects as risks with distinction between immediate, primary effects as "directly connected with the technology", and as "an inseparable aspect of its use" from the one hand, and from the other hand side secondary effects (indirect, consequential effects), "which spread by way of various and partially only insufficiently known case / effect relationships"; and "in general, the impacts and second-order effects of the direct effects of technology on the natural environment and on society" (Brune et al. 2006: 331).

The subjects of risk assessment hereby are the environmental or health risks, economic risks, risks of social conflicts and risks for sustainable development. Accordingly to these there are following types of technologically initiated risks: 1) Accidents in technical facilities, i.e. technologically caused catastrophes which have destroyed the social trust in technology so for instance in the case of nuclear power energy. 2) Consequences for human health such as completely new materials or emissions, e.g. risks and side effects of 'new' medicine or all the problems with asbestos and other hazardous substances. 3) Consequences for the natural environment such as air and groundwater pollution, the ozone hole, but these are often a progressive and gradual processes, so for instance in the case of GMO with anthropogenic irreversible containable modifications and changes of organic structures. 4) Social and cultural effects of technology, e.g. the impacts of a new technology on the labour market with the necessity to acquire new competences, to learn the new work-culture, but also all the impacts of converging technologies such as the new forms of technologically based social communication, or in the case of nanobiotechnology the new types of medical treatments. 5) Intentional misuse of technology above all in the case of terrorism as a new type of technologically unintended risk and the necessity to redefine of technological purposes. 6) Vulnerability of society with the dependence of the modern society from the technological infrastructure.

Therefore the risk and technology assessment shall be anticipative, i.e. 1) "spatially far-reaching effects up to global effects" (air pollution, degradation of natural environment, global water cycle), 2) "increase of temporal scope" of technological consequences, e.g. final disposal of radioactive wastes, 3) immense enlargement of the group of those possibly affected by hazards including the coming generations in the future, 4) "the problem of delayed effects" (so in the case of ozone hole), 5) "the difficulty of ascertaining the chain of causes in view of highly complex and hardly reproducible causal relations", e.g. the mad cow disease, 6) "poor or insufficient perceptibility of the risks (…) with normal human sensory organs", e.g. nano-pollution, 7) "diffuse distribution of responsibility" in the complex socio-political and technological system with society's vulnerability of technical dysfunction, 8) increased irreversibility of risk impacts, 9) and finally the lack of knowledge about the possible undesirable side effects. Concluding, the challenge by risk assessment "consists much rather in addressing the risks, in analysing them, and in evaluating them rationally, in comparing them with the expected benefits, and in taking the results of these deliberations into consideration in decision-making processes" (Brune et al. 2006: 333–334).

Concerning the ethical and societal impacts of nanotechnology two main subjects are especially worth to be considered, i.e. the precautionary principle and the life cycle approach concerning nano-based products and treatments. These two aspects of ethical and societal impacts of nanotechnology result above all from the risk assessment and the envisaged new framework of responsibility by dealing with science at all and with the emerging converging technologies. The risk assessment itself integrates or are focused on (cf. Allhoff and Lin 2009: 75): the protection of environment, the safe use and impacts on health, but also protection of values such as dignity and liberty of a person, the quality of life with economic and political impacts of nanotechnology, and connected therewith the new ethical controversies focused on privacy, justice and equity as expression of the ongoing process of science and research democratization basing on the principles of transparency and participation. The aim and goals of nano-risk-assessment are preparation of recommendations to the political decision-making process. Hereby the main research fields in nanotechnology are: nano-materials, nano-electronics and nano-devices, nanobiotechnology, nano-metrology. In all these fields are expected benefits form environmental protections to medical applications, from information and communication technologies to agriculture and food production. Following to these as pivotal with regard to the nano-development appears research on possible and hypothetical risks and side effects resulting from the nanotechnological convergence.

With regard to the technology assessment in the case of nano-domain important role play the degree of uncertainty and ignorance, i.e. the probability of side effects and the type of damages, that "they still raise questions of how to deal with scientific uncertainty, which in turn is related to the state of knowledge in this area as well as to ignorance about what could happen once a new technology is adopted" (Allhoff and Lin 2009: 77). The definition of risk as "possible, uncertain event that can cause damage and is not solely dependent on a person's desires" presupposes both the possibility of a potentially undesirable, unforeseeable and harmful event, and damage involved therewith: "These factors raise questions as to how to deal with scientific uncertainty tied to the state of knowledge in the field, as well as ignorance about what might happen after a new technology is implemented" (Allhoff and Lin 2009: 77). Concerning the monitoring of nanotechnology and nano-risk-assessment crucial are two subjects: the nanoparticles as condition of nano-applications and their impact on health and environment, and secondly the new materials and their properties with toxicity or reactivity.

### 4. Nanotechnology: risk assessment and precautionary principle

The task of risk assessment and management is to prevent the negative impacts and side effects on human health and natural environment. The main areas of risk assessment and management are: 1) working places and regulations concerning specific kind of exposures; 2) regulations in the case of nutrition and food; 3) elaboration of environmental standards of emissions, e.g. the groundwater quality; 4) standards of safe use and implementation of technologies and materials with regard to their toxicity. Risk analysis includes three elements: 1) risk assessment with hazard identification and characteristic, then the exposure assessment, 2) risk management, 3) risk communication basing on risk assessment and management concerning hazard, exposure and characteristics. These three components of risk analysis are interconnected and overlapped with the precautionary principle as their decisive framework and hazard identification as the starting point of analysis. Concerning nanotechnology it is the question how far the precautionary principle can be use "as a means of 'uncertainty-based' regulation": "In case that risk may not be calculated in this (quasi-objective) way there is uncertainty about the extent of harm and the probability of the occurrence of harm remaining. In this case 'uncertainty-based' regulation applies which, in case of possible severe harm, leads to the precautionary principle" (Brune et al. 2006: 335).

Following to this the risk analysis includes risk-based or uncertainty-based regulation. By risk-based regulation there are a set of successive steps: 1) identification of risk, 2) elaboration of normative standards including ethical analysis and public debate of acceptable risks, 3) definition of harm, 4) calculation of probability by side effects, 5) minimisation of risk and side effects as the objective of policy making process towards the protection of health and environment, 6) prevention of the long-term effects. – By uncertainty-based regulation the starting point is to underline the acceptable uncertainties and the normative transformable standards in the case of particular technology and/or materials. The aim is to assess "the plausibility of possible adverse effects", and the policy objective is the reduction of the scope of uncertainties with precautionary practice, i.e. "prospective long-term effects can only be identified to some extent, precautionary measures such as monitoring enable possibly the early identification of adverse effects over time" (Brune et al. 2006: 335).

Therewith the following assumptions concerning nano-risk-assessment can be distinguished: 1) the objective is to analyse the behaviour of nanoparticles, their impacts on animals, human and environment depending on the assimilation ways (lung or digestion); 2) the exposure to nanoparticles at working places, air in public sphere, consumers of nano-based products; 3) the specificity of nano-risk and the necessity of new regulatory regimes, i.e. how far nanotechnology is a challenge to the established risk analysis. These concerns above all the manufactured nanoparticles, which cannot be integrate in the classical risk analysis and classification system, because nanoparticles "might consist of well-known chemical substances or might fit into established classes of chemicals. But because of the new properties and functionalities they show and which are closely related to the nanoscale there might be new classes or types of risk and uncertainty in spite of the fact that the basic chemicals and their toxic properties might be well-known" (Brune et al. 2006: 336–337). Hereby the categories by the risk assessment in nanotechnology are: 1) the size of nanoparticles and their mobility and translocation in organism, 2) the morphology of nanoparticles, and 3) the surface properties with adhesion, cohesion, and agglomeration.

With regard to the nanotechnology and risk assessment appears the controversy with the scope of implementation of the precautionary principle: How and when it has to be used? The background of the controversy about the precautionary principle is "the dilemma of balancing the freedom and rights of individuals, industry and organisations with the need to reduce the risk of adverse effects to the environment, human, animal or plant health", where the undertaken actions should be "proportionate, non-discriminatory, transparent and coherent" (EC 2000: 3). The precautionary principle is an element of risk assessment, and especially risk management. With the dissemination of information and communication technologies and the increased flows of data and information is growing the awareness and sensibility of the potential risks and unintended effects resulting from new technologies on health and the environment by public opinion. In the first approach the precautionary principle can be defined as a "decision to take measures without until all the necessary scientific knowledge is available" (EC 2000: 8). In this manner precautionary principle is a tool "to help in decision-making when people face the uncertainty that so often characterizes complex ecological systems", and as a tool "for managing health and environmental well-being", and as "a guide to public policy decision-making under condition of scientific uncertainty" (Schettler and Raffensperger 2004: 64, 66).

Accordingly to the Wingspread Statement from 1998 "[w]hen an activity raises threats of harm to human health or the environment, precautionary measures should be taken even if some cause-and-effect relationships are not fully established scientifically" (Tickner et al. 2004: 1). This is the basic statement concerning legitimacy of implementation of precautionary principle, also as a more radical postulate to protect the environment and human health. One of the main arguments against the precautionary principle is above all the burden of scientific proof. In consequence the question remains how it is possible "to make preventive decisions in the face of uncertainty and to drive actions that will protect public health and the environment" (Tickner et al. 2004: 2). Hereby the case of asbestos can be regarded as the "detailed review of the use, neglect and possible misuse of the concept of precaution in dealing with a selection of occupational, public and environmental hazards" (Harremoes et al. 2002: 1).

Therewith the controversy about the precautionary principle underlines the political and societal background of use and dissemination of technologies, where precautionary practice makes out "a general rule of public policy action to be used in situations of potentially serious or irreversible threats to health or the environment, where there is a need to act to reduce potential hazards before there is strong proof of harm, taking into account the likely costs and benefits of action and inaction" (Harremoes et al. 2002: 4). In this manner the precautionary principle is expressis verbis the factor of science and research policy, and concerns "the political decision to act or not to act" and "how to act": "The precautionary principle is relevant only in the event of a potential risk, even if this risk cannot be fully demonstrated or quantified or its effects determined because of the insufficiency or inclusive nature of the scientific data" (EC 2000: 13). In the following considerations are presented the main constitutive elements of risk assessment such as characterisation, management and communication of risk in nanotechnology with regard to the precautionary principle.

## 4.1. *Characteristics of risk in nanotechnology*

In general, the risk makes out the integral part of innovation and technological development, that "without any risk, nothing will happen anymore", i.e. "no innovation without risk" (Brune et al. 2006: 339): innovation accompanied by risk-consciousness are the initiation points of technology development, but technology has always impact and side effects on human life and natural environment. In the case of nanotechnology the risk analysis and risk characterisation are focused first of all on the possible and unforeseeable toxicological effects. The starting point of the nano-risk analysis consists of identification of potential risks in the particular cases by consideration of the internal and external factors with respect to the environment, health and safety. In the nanotechnology risk assessment is to be distinguished between 1) the free nanomaterials (nanosized particles) and fixed nanoparticles with regard to their capabilities to translocate (differences in their mobility), 2) the nanoparticles produced in technological processes and nanoparticles produced unintentionally by conventional technologies such as diesel exhaust particles, fly ash, catalyst emissions or candle light and smoke-black, because the engineered nanomaterials with new properties can have totally different impacts on living organism with regard to their toxicity. These result from the plurality and diversity of substances, structures, surfaces and sizes, but also from their interactions with living systems. This underlines the significance of nano-risk-assessment at the moment of mass dissemination of nano-based materials and products.

The risk analysis of nanotechnology focuses on distinction between coated and uncoated engineered nanoparticles, where coatings change the characteristics of nanoparticles. The other question concerns the possible life span of nanomaterials, i.e. distinction between short- and long-lived engineered nanomaterials and nanoparticles, especially by their accumulation in the living organisms. By production and use of nanomaterials there are distinguished between the occupational and environmental implications, their distribution in the air and into the water. This concerns manufactured nanoparticles and nanomaterials for industrial and medical purposes.

First of all it is the distinction between intended (nanotechnology) and unintended (traditional technology) nanoparticles with regard to the new class of non-degradable pollutants appearing by nanomaterials production and use. Hereby the distribution in air, ground water and soil are the major ways of exposure with regard to the environmental risks (Colvin 2003b) with the question concerning the bio-accumulation processes of nanomaterials, how far they are non-biodegradable and can effect as persistent pollutant. Therefore four main challenges for eco-toxicological risk management by nanotechnological products are meanwhile distinguished: 1) the change of the toxicological research up to now, because of better understanding of the molecular and subcellular interactions with genomic and proteomic aspects; 2) the elaboration of models detecting toxicological side effects of nanoparticles and nano-based materials; 3) "linking molecular, cellular and pathophysiological 'endpoints' with higher level of ecological consequences"; 4) "precautionary anticipation of possible harmful impacts of novel developments in industrial processes, including biotechnology and nanotechnology" (Brune et al. 2006: 349).

Besides the environmental risks, the assessment of nanotechnological risk includes the analysis of health effects and impacts focused on identification and characterisation of possible hazards with the assumption of exposure and risk involved in. The health impact analysis bases on the dose-response relationship or correlation. The risks connected with these xenobiotic are expressed in toxicology with following terms and definition: 1) Hazard: "the potential of a noxious matter to cause harm: hazard is typically assessed by (occupational) toxicologists testing the harmful potential on cultured cells or isolated organs (in vitro) or directly on laboratory animals (in vivo)". 2) Exposure: "the presence of the noxious matter in a relevant medium (air, food, water, soil) multiplied by the duration of contact". 3) Risk: quantification of hazard and exposure (risk = exposure × hazard), i.e. "if a noxious matter is demonstrated to be harmful (hazard) and there can be situations defined where an exposure can occur (exposure), then there exists a possible risk for adverse effects induced by this xenobiotic"; the calculated risk depends on the correlation between the noxious matter and the time of exposure dose. 4) <u>Dose</u>: "the amount or concentration (e.g. for nanoparticles ng or number per m$^3$ air) of the noxious matter that will reach a specific biological system (e.g. ng/kg body weight)" and the dose depends directly on the exposure. 5) <u>Persistence</u>: "Chemicals or materials that do not decay or that either take a long time to decay are most threatening; because of their persistence the chance to burden living organisms will drastically increase". 6) <u>Bioaccumulation</u>: stability and persistence of chemicals and materials and their accumulation in particular organs (lipophilic xenobiotic and the accumulation within the food chain). 7) <u>Toxicity</u>: "concentration of a given substance or material in a given matrix that will induce direct effects on living organisms in that compartment". The categories of hazard characterisation are: the particles and their size, purity, technology production, the top down or bottom up approach, the type of substances (e.g. metal, carbon), open or closed use, and inherent safety (e.g. adhesion, aggregation). The routes of exposure are: ingestion, dermal penetration, and inhalation (Brune et al. 2006: 350–351). Following to this characteristics nano-toxicity is established as a new subdiscipline of toxicology.

The ways of nanoparticles by going into organism are: the digestive tract by ingestion, the respiratory tract by inhalation, and the skin by direct exposure. The special case is the iatrogenic exposure by injection of nanoparticles for medical purposes. Then, being in body, they may behave as xenobiotic and can translocate between organs, systems and tissues, so for instance by passing the blood-brain barrier and cardiac-nervous systems. The main way to entry into the body is the respiratory tract into the lung, and the main target organs by nanoparticles exposure are: 1) organs of intake such as skin, gastrointestinal and respiratory tract, 2) central nervous, cardiovascular and immune systems, and 3) liver. Open remains the question how far nanoparticles can be genotoxic with the ability to translocate into DNA.

One of the first recommendations devoted to the life-cycle assessment of nanomaterials with potential risks is elaborated by European Commission (EC 2004a). This life-cycle assessment of nanomaterials includes: exposure events and uptake, then bio- and toxico-kinetics and toxicology of nanoparticles and nanomaterials, their risk characterization and communication, but also risk-mitigation, finally their emission as transport and transformation in air, water, soil and plants. European Commission (EC 2004c) underlined the most important issues concerning life-cycle assessment on nanotechnology based products: 1) the identification of safety issues at all stages of use and production of nano-based products, 2) the support of nano risk assessment in the research work and R&D programs including manufacturing, distribution, use, and disposal,

3) the elaboration of data concerning nano- and ecotoxicology and evaluation of the possible human and environmental exposure. Finally, with the first attempts of the military use of nanotechnology a revision the arms control agreements (e.g. Biological Weapons Convention), or the international law of warfare with introduction of a new element such as autonomous fighting systems and with the new types of terrorism in the future, are needed (Altmann 2004).

*4.2. Risk management in nanotechnology*

The risk management of nanoparticles and nanomaterials consists of hazard's identification and characterisation on the basis of dose-response exposure assessment. But there is a lack of data concerning dose-response exposure in the case of engineered nanomaterials. The goals of risk management are: the constitution of framework of nanomaterials regulation, the applicability of precautionary principle and the elaboration of regulatory measures. The framework of risk management with a view to the regulation issues includes four steps of risk assessment: 1) Hazard identification with multiple effects of nanoparticles appearing in organs of intake and secondary organs through in vitro (cellular system) and in vivo (animal system). 2) Exposure assessment with distinction of the exposure routes with biological monitoring (markers of exposure) and occupational and environmental monitoring. 3) Dose-response assessment basing on hazard identification from the one, and occupational and environmental monitoring in the exposure assessment from the other hand in in-vivo studies (acute, chronic, and species dependent) and in-vitro studies (primary cells, cell lines, co-culture-systems). 4) Risk calculation basing on exposure assessment (biological, occupational and environmental) and dose-response assessment with data from in vivo and in vitro studies. In this manner risk calculation includes biological mechanisms analysis and extrapolation of susceptibility (high and low, animal and human). These steps of risk management, finally basing on risk calculation, should lead to the elaboration of a set of risk prevention and minimisation measures, what makes it possible for application in form of occupational and environmental regulations with exposures values.

The public debate on nano-risks started in US in the 1990s with the first regulatory approaches by US Food and Drug Administration basing on the Toxic Substances Control Act (TSCA) as US Regulatory Act for Chemical Substances from 1976 and the TSCA Chemical Substance Inventory (cf. Haum et al. 2004: 40–44 in the case of nanotechnology). In these pragmatic and product-based approaches the precautionary principle was not used with the argument, that a regulatory action "can only be undertaken if any scientific uncertainty about possible hazards has been resolved. The 'burden of proof' applied requires that there is clear scientific evidence for the existence of adverse effects" (Brune et al. 2006: 371). In the case of nanotechnology there is distinguished between three exemptions: 1) the low volume exemption, i.e. regulation by production about 10 tons per year of a particular substance, in the EU 1 ton accordingly to the regulation of chemicals REACH (cf. Haum et al. 2004: 45–49); 2) the low release and exposure exemption; 3) the test marketing exemption. Besides US regulatory systems, European REACH system and British studies (RS/RAE 2004) the first recommendation is elaborated by European Commission (Community Health and Consumer Protection, cf. Haum et al. 2004) with classification of nanomaterials with regard to such categories as risk, toxicity, and proliferation.

The most important controversy in the case of nanotechnology risk assessment and regulation concerns the application of the precautionary principle. The traditional risk regulation is "adequate if the level of protection is defined, and the risk (e.g. the probability of the occurrence of the adverse effects multiplied by their impact) can be quantified. In such situations thresholds can be set, informed by ethical analysis because of their normative nature (…), risk can be either minimised or kept below a certain level, and also precautionary measures can be taken to keep particular effects well below particular threshold" (Brune et al. 2006: 372). This traditional risk regulation bases on the ALARA principle, i.e. as low as reasonably achievable, and results from scientific information, empirical data and knowledge, and can be used by appeared uncertainties (cf. EC 2000: 15). But this situation changed radically by the converging technologies. "This is notably the case when epistemic debate is going on in science: different disciplines use competing models or analogies or basic assumptions to disclose the subject matter under investigation in order to acquire new knowledge. As soon as the conditions required for the risk management approach are no longer fulfilled, controversies and ambivalent situations under uncertainty are the consequence"; and this is valid in two cases: 1) "if scientific knowledge concerning possible adverse effects is not available at all or is itself controversial and hypothetical, or if empirical evidence is still missing (for example, if the knowledge is based on models which could not yet be validated to a sufficient extent)"; and 2) "if adverse effects could have catastrophic dimensions with respect to their amount or to their specific impacts and consequences", and where the 'future ethics' with the principle of responsibility could be implemented (Brune et al. 2006: 373; Jonas 1979).

The cases of asbestos, mad cow disease and the ozone hole initiated the public debate about risk assessment and management, and underlined the postulate of stronger regulation measures application in the field of new technologies. Above all there is presented the idea of precautionary principle implemented for instance in the EU. At first time the precautionary principle was introduced in the Rio Declaration from 1992 as a part of Agenda 21 with the principle no 15. The precautionary principle is used above all in the field of global climate change, ozone depleting substances and conversation of biodiversity. In the EU Treaty of Maastricht from 1992 the precautionary principle is included in the article 174 as the base of the European environmental policy (cf. EC 2000). The characteristics of precautionary principle includes the measures or categories understanding also as integral part of risk management, such as: 1) proportionality that actions shall be adequate to the desired level of protection; 2) non-discrimination that actions shall be adequate to the situation; 3) consistency with other already adopted measures and scientific data; 4) balance of costs and benefits including economic factors but also the public opinion; 5) reviewing of the measures accordingly the monitored development of science and research and the changing scientific data with examination of scientific development; 6) assigning of responsibility, the producer shall demonstrate the danger, risk and safety paths of use. In this manner precautionary principle "combines the ethical notion of duty to prevent harm with the realities of the limits of scientific understanding" (Schettler and Raffensperger 2004: 66). And the three core elements of precautionary principle are: 1) identification of harm, 2) dealing with uncertainties, and 3) understanding of precautionary as action. The key element hereby is the uncertainty (Tickner et al. 2004: 12–14) with the following categories: 1) statistical or parameter uncertainty by missing information and scientific data, and with the value of a single variable; 2) model uncertainty with more than one variable interacting in a complex system, but with finite number of variables and their interactions; 3) fundamental, systemic or epistemic uncertainty, especially by long-term or large-scale analyses, which expresses the indeterminacy of a complex system and resulting from this ignorance. Therefore the uncertainty is a permanent aspect of precautionary principle: "Judging what is an 'acceptable' level of risk for society is an eminently political responsibility. Decision-makers faced with an unacceptable risk, scientific uncertainty and public concerns have a duty to find answers. Therefore, all these factors have to be taken into consideration" (EC 2000: 4).

With regard to the political process of decision-making there are following conditions of implementation of precautionary principle: 1) establishment of "the level of proof needed to justify action to reduce hazards (the 'trigger' for action)"; 2) the early detection of hazards by research and monitoring; 3) "the proportionality principle, which states that the costs of actions to prevent hazards should not be disproportionate to the likely benefits"; 4) undertaking of "action to reduce risks before full 'proof' of harm is available if impacts could be serious or irreversible" (Harremoes et al. 2002: 4). And the guidelines for applying the precautionary principle are: 1) scientific evaluation of risks including the level of scientific uncertainties; 2) "assessment of the potential consequences of inaction" (EC 2000: 17); 3) respecting the principles of risk management such as proportionality, non-discrimination, consistency, balance of costs and benefits, evaluation of scientific development (cf. Harremoes et al. 2002). Finally the tools of precautionary policy are (Tickner et al. 2004: 5–6): 1) bans and phase-outs as the most radical action; 2) clean production and prevention of pollution; 3) alternative assessment; 4) introduction of occupational exposure limits; 5) ecosystem management concerning the maintaining of biodiversity with underlying importance of reversibility of the interventions in the ecosystem.

The application of precautionary principle starts with scientific examination of a material or technology, i.e. which uncertainties (theoretical and practical) and risks are involved in, from the one hand, but from the other hand "in the case of epistemic uncertainty, a normative relationship between the nature of the uncertainties and the possible adverse effects needs to be established in order to justify policy and regulatory intervention", and with regard to the nanotechnology the question is: "Is there 'reasonable concern' (…) to legitimate the application of the precautionary principle? If yes, what would follow out of this judgement with respect to adequate precautionary measures?" (Brune et al. 2006: 375, 376). But the obstacle by nano-risk-assessment is "the general lack of high quality exposure and dosimetry data both for humans and the environment", so that "there is not yet a generally applicable paradigm for nanomaterial hazard identification, a case by case approach for the risk assessment of nanomaterials is warranted" (EC/SCENIHR 2009: 8, 4).

The necessity of regulation and the dilemma by application of precautionary principle in the case of nanotechnology refer to the responsibility towards the uncertainties about nano-risks. Accordingly to the precautionary principle the question is "whether there is, according to present knowledge, 'reasonable concern' about possible adverse effects of nanotechnology which could serve as legitimating reason for very strict measures (like a moratorium on nano-particle use in products, a strict regulation of the release of nanoparticles or even a ban on nanoparticles)". Hereby the condition of implementation of the precautionary principle is a scientific assessment of the state of knowledge, but also the quality of the available data as the base of normative analysis with regard to the responsible development. In the case of nanotechnology "there is no reason for serious concern", i.e. there is, "with regard to nano specific effects (excluding self-organisation effects and cumulative effects of mass production), no reason for particularly great concern about global and irreversible effects of the specific technology per se" (Brune et al. 2006: 376–377).

But, confronted with the lack of knowledge and data, the question still remains open how far the precautionary principle shall be used also in the case of nanotechnology and the different fields of impacts: "Scientific uncertainty is, following the precautionary principle, not a sufficient reason for strict measures though the situation cannot guarantee that there will be no serious harm implied by a wider use of nanoparticles. It will always be possible to create speculative scenarios, however, does not legitimate to use the precautionary principle as argument for a moratorium or other prohibitive measures" (Brune et al. 2006: 377). That means also a considerable difference regarding to the technological conception, so for instance the future ethics by Hans Jonas (1979), where hypothetical assumptions make part of normative evaluation. The alternative to the precautionary principle are the self-organisation of science basing on the principle of responsibility, the adaptation of existing regulations to the case and specificity of nanotechnology, and the development of systematic and permanent monitoring framework of nanotechnology focused on the possible nano-risk indication as a kind of preventive attitude comparing for instance to the asbestos-case. The first implementation of precautionary principle in form of substantial precautionary framework is represented by the case of GMO with the Directive of European Commission from 2001, although there was no scientific or experimental evidence for harm to the humans or the environment caused by GMOs. The precautionary principle is translated in this case into a regulatory framework with case by case procedure. This case by case procedure concerns the microscale of particular GMOs, nanoparticles or nanomaterials regarding to safe use, production and dissemination. The precautionary principle would imply "that there is a need for a cautions step-by-step diffusion of risk-related activities or technologies until more knowledge and experience is accumulated" (Haum et al. 2004: 6).

The relevant issues by the implementation of precautionary principle in the case of nanotechnology are: 1) normative aspects of designing regulation under uncertainty; 2) technology assessment; 3) evaluation of potential effects. And with regard to the nano-applications there shall be distinguished between: 1) the 'dry' applications concerning the nanoparticles with the potential unforeseen adverse effects on health and environment, and 2) the 'wet' applications in the case of convergence of nano- and biotechnologies. With respect to the designed nano-development in the future the precautionary principle can be used in form of a characterization of this technology which could be regarded as a base of future action. But form the other side the development of science and technology is characterized by uncertainty, so that instead of precautionary principle it shall be developed science-based framework of risk assessment with respect to the establishment of a more rationally based approach by the governance of nanotechnology and the political process of decision-making, because the precautionary principle is "too vague and too arbitrary to form a basis for rational decision-making", but at the same time "proponents of precautionary regulation point out (…) that there is a need for a cautious step-by-step diffusion of risk-related activities or technologies until more knowledge and experience is accumulated and a risk based regulations may have shortcomings against a background of scientific uncertainty" (Haum et al. 2004: 8). So that the controversy in the case of nanotechnology oscillates between the precautionary principle as the key tool based on uncertainty or risk assessment as more rational approach. From the one hand the argument is: "If technologies are to be designed and developed with a view towards both safety and sustainability, it is essential to carry out technology assessment at an early stage and to understand the different types of innovation process involved" (Haum et al. 2004: 17), and from the other hand the fact is that here is "no life without risks. There is no way towards a sustainable economy and society without innovation since innovation and risk are inextricably linked" (Haum et al. 2004: 44).

Concluding, there is needed a completion between precautionary principle, risk analysis and principle of responsibility, and the main issues hereby are: 1) application of regulation of chemicals to the nano case with treating nanoparticles as new chemicals as the base of risk management accordingly to the existing codes of conduct and regulation measures; 2) respecting the lack of knowledge, especially in the case of long term (side) effects of nanoparticles on health and environment; 3) with regard to the nano-domain there are not a totally new situation from the ethical point of view, and the main challenge is the scientific development of nanotoxicology focusing on the side effects by the humans and the environment, what underlines the necessity of a synopsis basing on comprehensive and permanent evaluation and monitoring of development and implementation. Following to this nanoparticles and new nanomaterials shall be treated and classified as new substances accordingly to the chemical substances with three mains categories: risk, toxicity and proliferation. 4) Instead of a nano-moratorium is needed the establishment of institutional monitoring framework of nanotechnology, but also with integration of a dialogue between public opinion, scientists, research centres and industry. This dialogue should contribute to the establishment of comprehensive evaluation of the state of knowledge, especially in the case of nanotechnology as a converging technology being in permanent state of changes and development. This case by case strategy makes it possible to implement realistically the precautionary principle in the case of nanotechnology: "The present situation 'no reasonable evidence of harm' does not mean 'evidence of no harm'. This situation implies that science and industry should deal with nanoparticles (in research and industry) in a cautions way, due to their responsibility and accountability" (Brune et al. 2006: 379; cf. the following analyses of nanotoxicology in Part C).

The main challenge for risk assessment is the high complexity of the human/ societal and ecological systems with a set of multiple and various changing and interacting variables: "Risk assessment, which was originally developed for mechanical problems such as bridge construction, in which the technical process and parameters are well-defined and can be analyzed, took on the role of predictor of extremely uncertain and highly variable events" (Tickner et al. 2004: 14). In consequence the risk assessment as a tool can only manage and reduce but not prevent the risks. The objectives of risk assessment are quantification and analyses of risk-problems but not to solve them. Focusing on 'acceptable risk' the risk assessment can be regarded as expression of industry but also policy driven science, and therewith undemocratic and anti-participative with respect to public opinion. In consequence the collective responsibility has replaced the responsibility of those who created harm. Moreover, the "precautionary approach asks how much harm can be avoided rather than asking how much is acceptable" (Schettler and Raffensperger 2004: 66), what makes also the distinction toward the risk assessment. By risk assessment the objective is a particular product, and the assessment is focused on statistical and quantitative single variable, in consequence the model uncertainty and the fundamental uncertainty are interpreted as the statistical one. In consequence risk assessment "failure to predict adverse effects removed in time and space and complete failure to predict surprises or novel effects" (Schettler and Raffensperger 2004: 76), in end-effect the risk assessment is a risk minimization which bases on the costs-benefits analyses. Precaution deals with complex systemic interactions and multi-causal dimension with the distinction of the different types of uncertainties, what means an assessment of relative risk. The guideline of precautionary action is the protection of resilience and diversity, but at the same time the precautionary approach bases more on the visions. The dilemma between risk assessment and precautionary principle is relevant with regard to the nano-visions and public communication with the question how far precautionary action are nano-visions' driven as for visions-based public opinion about nanotechnology.

## 4.3. Strategies by nano-risk communication and nano-visions assessment

With the assumption that the development of new technologies depends on societal acceptance and the possibilities of public funding of research involved in, one of the strategic fields in the advance of science and technology is the

## B. Nanotechnology and Development of Assessment Regime 135

public communication. The main issue or background of the interconnectedness between science, technology and society is above all the changed relationship between university, research and technology transfer with industry and private sector, and between basic and applied sciences. The driving forces of this change are the efforts to stretch the time by passing from labs discovery into the marketplace, from prototype into product. The traditional understanding of these relationships with the model of mission-oriented science is questioned in the last decade. The expression hereby is the US National Nanotechnology Initiative where it is a question of "the next industrial revolution, not the next scientific revolution" (Roco and Bainbridge 2007: 81).

Besides the changed relationship between science, technology and society the institutional changes in research and technology transfer has modified the understanding of public and private goods, and therewith connected the intellectual property especially at the institutions of higher education: "When one begins to think seriously about molecular manufacturing, intellectual property becomes that much more important. These changes are having profound implications for the distribution of risks and benefits that come with new technologies" (Roco and Bainbridge 2007: 82). Nanoscale and molecular manufacturing express the ongoing research at the interfaces between nanotechnology and biotechnology. Hereby nanotechnology can be compared with biotechnology and the entering of genetically modified foods and products into marketplace, and the public debate and controversy which are related therewith. In this manner appeared in the last period a new field of research – nanobiotechnology, which remains a challenge concerning public communication about its risks and benefits, and underlines the necessity of permanent monitoring of nano-development from societal and ethical point of view, because of the ongoing research, accelerated development and permanent change the state of affairs. The objective is not to develop speculative scenario in a long-term perspective, but the analysis situated 'in-time' of the present ongoing research and outcomes. The societal and ethical assessment has to accompany the scientific and practical (applicable) development of nanotechnology. From the other side the nano-assessment shall include a prospective orientation: "But we are in a good position to establish social habits and institutions that are able to consider nanobots if and when they arrive" (Roco and Bainbridge 2007: 82) with three time frames: 1) now, 2) near-term or medium term about 3–7 years, and 3) long-term about 10–20 years.

By the medium term (2007–2015) the objective of nano-assessment are the manufactured nanoparticles entering into the market with estimation of their toxicity to humans and the environment. By the long-term time frame (2010–2020) it is for instance the question concerning the 'interactivity' of nanoparticles in living organisms, their capability to remove between the different organic systems (from cardiac to nervous) and their medical applications, and finally the new materials and their properties such as graphene. The last and newest problem concerns the treatment (recycling) of used nano-based products, and construction of the prototype nano-devices. The other questions concern privacy and human enhancement. The long-term assessment focuses above all on the new type of interfaces between human and nano-devices, and the possibilities of molecular manufacturing.

Technologically caused risk communication includes the relationship between science, technology and society, and therefore it shall be considered "that risk perception in society and public risk communication did not simply follow the scientific assessments but rather went their own course, frequently leaving back confused scientists who could no longer understand what was being going on" (Brune et al. 2006: 380). As examples we can mention the case of nuclear power and genetic engineering with enormous problems with communication between the community of scientists and the public opinion, which have determined both development and implementation of these technologies. In consequence, in the 1980s is elaborated a new approach with risk research in social and political sciences with the objectives such as public risk perception and patterns of risk communication to the public opinion. In this sense emerges the importance of messages concerning nanotechnologies in the mass media, i.e. creation of public and social attitude towards nanotechnology without the 'risk' to identify nanotechnology with the negative images associated with nuclear technology or genetic engineering.

Nanotechnology became a subject of public debate in the 1990s with a positive image, but this changed with the paper *Why the future doesn't need us* by Bill Joy (2000) with a horror scenario about nanotechnology and its ambivalence. In the followed time the public debate about nanotechnology has been dominated by visionary and science-fiction scenarios such as 'grey goo', 'nanobots' or 'cyborgs', also as motifs of the mass media and pop culture. In this manner nanotechnology is anticipated as a risky technology. Hereby, the futuristic visions in public debate concerning technological development play an important role as well as in the development and dissemination of a technology and in the communication between scientists or engineers and the public. The communication patterns should increase public awareness on technology development but also the acceptance for public funding. But the presented hitherto nanotechnological visions, so for instance the first one by Richard Feynman (1960) and then above all by Eric Drexler (1986), extrapolated the specific risks of communication to the public as for the emerging new technologies.

## B. Nanotechnology and Development of Assessment Regime

The best example of ambivalent visions on nanotechnology is the 'grey goo' fatalistic doomsday scenario with nano-scaled and self-replicating robots which are out of control. But this vision or utopia presented by Drexler has had consequences to formulate for instance the patterns of risk assessment such as: 1) the technological assessment identification of each nano-based products or the differentiation between passive (e.g. passive surfaces) and active (e.g. 'robots', or rather 'assemblers', in the human cell) nanotechnology products; 2) the postulate of continuous monitoring of production and use of active nano-based products but also their life-cycle monitoring; 3) labelling of nano-products; 4) introduction of the participative framework integrating different social groups of interests and the public opinion. The 'grey goo' scenario is characteristic for the popular assessment of scientific and technological progress in the second half of the 20$^{th}$ century with nuclear technology, genetic engineering and today with converging technologies. But this scenario expresses also the critical attitude towards technology with the postulate of cautiousness and responsibility by development, use and mass dissemination of technology. – In the same way as 'grey goo' scenario, the so called 'prey' scenario presents a vision of the self-replicating nano-robots in the future. At first time it was a positive utopia with new possibilities of enhancement the human condition of life, especially in the medicine and the new medical treatments. This changed with the paper presented by Bill Joy (2000) with the argument, that converging technologies lead us to the reality of uncontrolled intelligent nanorobots understanding also as equivalent and completion of the genetic engineering. This losing control over technology remains a classical motif of the apocalyptic and utopian thought concerning the relation between humans and technology, and shapes the patterns of philosophical and societal reflection about relationship and attitude of the man confronted with technologized civilization, so that the "more capabilities could be given to technology the more fears in this respect arise" (Brune et al. 2006: 384). The third dominant scenario, the 'cyborg' scenario, focuses on the nano-biotechnological 'cyborg' and integrates the advanced research on the molecular biology level, nanoscale and engineering technology, above all with the new generation of nano-electronics and neuroimplants (neurobionics) which enable new direct interfaces between technological devices or artefacts and the human nervous system, also as the enhancement of human capability. In this sense the idea of convergence of humanity and technology is appeared.

The background of the visions' assessment, what underlines R. Hanson in essay *Five nanotech social scenarios. A contribution to visions' assessment* (Roco and Bainbridge 2007: 109–113) is the development of the information and communication technologies with the distinction between plausible and implausible claims about Internet and ICT. In the case of nanotechnology and its social implications in the vision presented by E. Drexler (1986) appears also a design of a new pattern of economy and manufacturing. The fact is the process of coalescence of new research fields after the first period of nanotechnological development, what is also assumed in the first visions with nanotechnology as a technology of new type of devices, the 'assembler', with totally novel and general capabilities which are in the same measure revolutionary like computer technology (cf. Baum 2003 in context of Drexler-Smalley debate). Meanwhile there are distinguished five economic and social scenarios of nano development and use from rather conservative to radical and futuristic point of view: 1) Scenario of atomic precision with the underlined possibility to operate with the atoms at the nanoscale and atom-scale manufacturing; this possibility is recognized as condition of many new products such as computers, sensors and medical implants. 2) General plants scenario with the manufacturing of general plants and the 3D-printing or direct manufacturing, first of all connected with the laser-based additive manufacturing technology (Amato 2003) with the model of production nearby customer. 3) Scenario of local production which is characterized by high level of generality and automatization and makes possible a self-production on demand with elimination the costs of transportation and labs. 4) Over capacity scenario with local general plants as a kind of mass consumption which become idle like PCs today. 5) Self-reproduction scenario, i.e. "a local manufacturing plant can create a copy of itself": "where plants built to atomic-precision can reproduce themselves (as all life forms do today), and in addition be programmed to produce other products (as a few life forms can now do in some limited ways)" (Roco and Bainbridge 2007: 112); this scenario can be seen as the most radical and futuristic concerning nanotechnology development.

In this manner such scenarios like 'grey goo' or 'prey' appear as the dominant element in public communication and debate about nanotechnology, what underlines at the same time the necessity to integrate the scientific perspectives in these debates and communications. This is also the objective of visions' assessment basing on the present state of knowledge concerning technological development, so for instance the efforts to create an artificial brain remain not only expression of utopian thought but realistic research design with a set of philosophical and ethical challenges. In this manner nanorobots or nano-biotechnological cyborgs appear as the new imagination of the modern Golem.

Nowadays the futuristic (long-term) scenarios concern particular research fields and innovative technology such as stem cells, brain research and nanotechnology, and are focusing on the development of a technology (scientific-technical visions) from the one hand, and its impacts on human and societal development from the other hand. Mixed together they base on assumption that nanotechnology is a symbol of the 3$^{rd}$ industrial revolution. In the case of vision assessment as a part of technology assessment the aim is to introduce a more critical, rational, reflecting and transparent framework for a prospective assessment, which could be a kind of prevention by confrontation with unforeseeable and undesirable side effects on humans and the environment. The objectives of visions' assessment are the normative contents and the potential moral conflicts as a kind of ethical vision assessment itself. This underlines at the same time the significance of visions and innovations by technological development. First of all innovation is the condition of a technology: "Technologies emerge with successful innovation. (…) A 'disruptive technology' emerges first by the establishment of a new product in a niche market. Thus, by examining the appearance of newly established niche markets, the path of change may be anticipated" (Roco and Bainbridge 2007: 127; cf. Christensen 1997). Following to this a 'vision' prepares the first possible impact and implementation of an innovation, and this concern also the development of nanotechnology. Pronouncements extolling the possible impact of nanotechnology to be far beyond what is possible have provoked criticisms of the science in proportion to the exaggerated visions enunciated, so in the case of the treatise by Bill Joy (cf. Binks 2003), or the book *Prey* by Michael Crichton (2002). Generally, "[w]e have to distinguish between the degree of facticity of the content of the visions and the fact that they are used in real communication processes with their own dynamics. Even a vision without any facticity at all might cause real impact on debates, on opinion-forming, on acceptance, and even on decision-making", that they can be "instrumentalized to create public acceptance and political support" (Brune et al. 2006: 424), so in the case of the NBIC initiative in the US. These visions (positive and negative) have impact on political and public debate, so that the necessity of their assessment is obvious.

The first category of technological vision assessment is the '*Leitbild* assessment' which is "more technical in nature and close to ongoing technological development (…), more or less 'realistic' and do not cover such a large time frame"; the second category are the long-term futuristic visions, which integrate both, "utopian aspects as well as the claim for feasibility by scientific and technological means" (Brune et al. 2006: 425, 426) as a specific hybrids of facts and fictions. The categories of the visions are: fantastic, futuristic, *Leitbilder* vision, and concrete goals and aims setting for technology development strategy as integral part of research policy.

The fantastic vision as expression of imagination contains any realistic elements and has a low degree of facticity. The futuristic vision is a long-term, technological and societal prospect with claim for feasibility, but at the same time with high degree of uncertainty, where facts and fictions are mixed together; the supposed aim is the enhancement human performance, what is characteristic for the technodeterminism, i.e. technology as the only drive-force of development, at the same time they are normative and predictive, so for instance the assumption that converging technologies are milestones of development and need a roadmap for achieving the aim (Roco and Bainbridge 2003), and that means a realisation of vision such as by Drexler. The 'Leitbilder' vision or guiding vision descripts the technical, planed technological future. The strategic vision concerns concrete goals and aims of technological development as base of research policy and programs, which is plausible and rational. There are positive aspects by all mentioned vision-types, they function as base of scientific and technological progress, and are regarded as motivation of societal, political and educational agenda development. – The analytical steps by visions assessment are: 1) the substantial part with content of the vision, 2) the pragmatic part using in communication, 3) the evaluative part with judgement of the vision's content, 4) the managerial part, i.e. how to deal with visions. The futuristic visions have a hybrid nature (facts and fictions), uncertainty, ambivalence (including positive as well negative aspects, positive or negative visions in the same measures), risk of backlash, i.e. too much promises in too short time. – At the same time the question is how far a 'Codes of Conduct' is for vision communication possible or needed, especially by visions assessment in the field of nanotechnology: "The more speculative and revolutionary the vision, the more interpretations are possible – a wide field for public debate and occasions for ideological use of them. Against this background, it is not surprising that the public risk debate emerged from the conversion of positive utopia to negative dystopia" (Brune et al. 2006: 430).

The question is how far these nano-visions are scientifically relevant. All these scenarios are basing on the technological idea of molecular manufacturing and the possibilities of self-assembling or self-replicating of autonomous nano-units or nanomachines. But till now there is no evidence of the possibility of mechanical self-replication. This was also the topic of the discussion between Eric Drexler and Rick Smalley (co-recipient of the Nobel Prize for Chemistry 1996 for the discovery of carbon 60). Smalley used in this context by expressing his objections the terms 'thick fingers' and 'sticky fingers' (Smalley 2001; Baum 2003). The question of the self-replicating potentiality is negative with regard to the mechanical processes, but it still remains open on the point of junction between mechanics and biology. In consequence, the mechanical self-replicating nanomachines, and the 'grey goo' vision are expressions of science fiction. At the same time, the case of neuro-nanotechnologies underlines the relevance of the 'cyborg' scenario as a form of convergence between technological system and devices and the human nervous system. In this manner the nano-informatics and nano-electronics begin to be integrated in the natural systems, but: "It seems to be rather probable that traditional borders between humans and technological systems will be transgressed in the next decades, and new interfaces will be established. The 'cyborg' scenario only points to the possible negative consequences of such developments" (Brune et al. 2006: 387). The new interfaces between human and technology are above all the new field of research in medicine, and in this sense nanotechnology is a challenge for society and human self-understanding. The negative visions concerning particular technology developments underline the need for vision assessment itself, because "occupation with futuristic visions, speculations and risks is not only sensible, but necessary because of their impacts on really ongoing societal discussions", but there still remains the fact, that "a statement that they would be completely impossible in the future, cannot be proven – new scientific knowledge might change our feasibility judgments" (Brune et al. 2006: 388; Grunwald 2004). This concern the long-term and unforeseeable today development of science and technology, but from the other side an active participation of the scientists by visions' assessment with respect to public opinion about technology advance and the risk perception is indispensable.

Nevertheless there remains the risk of visions' communication in public debate. In public communication about nanotechnology there are two aspects: 1) the ambivalence of positive visions of nano-development and 2) the frustration issue. The ambivalence of nanotechnology results from the assumption that nanotechnology is the base of the 3$^{rd}$ industrial revolution, and as a part of converging technologies contributes decisively to the enhancement of human performance by new interfaces between brain-brain human interaction, and between human, natural and technological systems (Roco and Bainbridge 2003: 1) Following to this statement nanotechnology as a completely new technology consists also of ambivalence, because: "Traditional convictions and values could be challenged and existing societal structures could change radically. There might be, in the course of time, winners as well as losers, there might be unexpected and possibly negative consequences, and, in any case, there will be a large degree of uncertainty" (Brune et al. 2006: 389). At the same time nanotechnology means also new fears. Finally the ambivalent character of nanotechnology concerns the public opinion itself, which oscillates from fascination to detection of the dark side. Therewith is related the frustration issue which results from the expectation of concrete results, it is the difference between the time of implementation and mass dissemination of scientific and technological innovation and public expectation concerning new products, devices, materials etc., so for instance by the space discovery confronted with the 'real' benefits and losing the public funding for the research development. The question is how far nanotechnology is overselling and is confronted with disappointments resulting from exaggerated expectations, so for instance in the case of personalized nano-medicine.

Concerning risk communication shall be distinguished between scientific risk assessment and the public perception of risk. Especially public perceptions of risk can determinate development of nanotechnology: "Innovation and progress are inevitably related to risk but also inaction carries risk. There is no choice between risk and no risk but only between different rations between risks and benefits. This simple message must be a 'ceterum censeo' part of the communication between science and the public" (Brune et al. 2006: 396). That means: 1) integration of realistic risk in public debate; 2) introduction of the transparency principle with communicative and participative instruments of technology assessment as framework of risk public communication; this is also the new task of the scientists and engineers to learn basic knowledge about risk perception, management and communication; 3) the principle of long-term monitoring of research as an open process because of incompleteness and uncertainty of knowledge.

## C. Ethical Aspects: Nano-Safety and Nanotoxicology

With the risk analyses are directly related the ethical aspects of nanotechnology with the question concerning the responsibility of unforeseen side effects, i.e. the toxicity of nanoparticles and nano-based products. Following to this the complex assessment of nanotechnology focuses also on the practice of the human researchers, that "a wide range of questions and problems which by no means are scientific or technological ones, since they do not concern atoms and molecules, machines and organisms, but the practice of the human researcher" (Brune et al. 2006: 28). At the same time the ethical and then societal aspects of nanotechnology accordingly to the assumptions of technoscience shall be enlarged by integrating besides the humans also the non-humans laboratories' infrastructure. The main objectives of the ethical analyses of nanotechnology are the health and environmental, but also societal and political impacts, basing on balance of risks and benefits. The classical risk analysis and assessment is not applicable viewing the possible hazards resulting from use and implementation of the manufactured nanoparticles, is argued from the one side, so that there is a necessity to elaborate a new risk assessment regime. By the missing data of possible hazard and harm of nanotechnology implications remains only the use of the strict precautionary principle, especially in the field of nanotoxicology. But this would mean to stop the research and development in a situation where a 'reasonable concern' in favour of implementation of a strict moratorium is not given. In this situation of uncertainty is formulated the postulate to handle with nanoparticles in the same way as in the case of new chemicals, so the argument from the other side.

This is also the objective of nanotoxicology as a new subdiscipline of toxicology, because "it has been clearly demonstrated that the degree of toxicity of nanoparticles is tied to their specific surface and resulting new properties, not their mass" (Allhoff and Lin 2009: 78). Moreover, nano-products change the understanding of exposure, i.e. "the fate, persistence, and bioaccumulation of products in the environment or in living organisms", and then the intensity of the side effects (Allhoff and Lin 2009: 79; EC/SCENIHR 2009), so that in the case of nano-impacts on health and environment a new regime of risk assessment is needed. The method of nano-risk-assessment includes: 1) the precautionary principle as a principle of action, and 2) the life cycle approach. The precautionary principle refers to the 'danger' as exposure to possible harm from the one hand, and to the 'risk' as possible suffering from uncertain danger from the other hand. The risk can be known or hypothetical, supposed one. Following to this "the possible occurrence of danger is what creates a risk" (Allhoff and Lin 2009: 81). The other distinction underlines the unreasonable and irreversible risk from the one side, and the danger which could be morally unacceptable from the other side. But the main problem from scientific point of view is to define the 'unreasonable' risk. The other dimension of risk concerns social communication, i.e. risk perception and acceptability by public opinion accordingly to the participative and deliberative principle by risk management[37]. But this results from the state of knowledge characterized by the degree of certainty, uncertainty and ignorance with regard to the risks. Finally the dealing with risk can be described by prudence confronted with the known risk, then prevention in the case of uncertainty, and precaution by hypothetical risk: "Instead of seeking to avoid all risk, the question is probably how to determine which risks are acceptable. What adverse effects might a risk have? What is their scope and magnitude? What is the probability? (…) What avenues are possible to ensure that the precautionary principle leads to action?" (Allhoff and Lin 2009: 82). Precaution as a principle

---

37 Cf. the distinction between them presented by Bińczyk, pp. 42–48.

for action respects the principle of responsibility anticipating unproven and hypothetical risks as subject of debate in a pluralistic and democratic society.

In the case of nanotechnology assessment the objective of the ongoing research is the construction of nano-devices which can be used in toxicology with regard to elaborate standard procedure accordingly to the assumptions forming the precautionary principle. First of all the constructions of nano-devices make it possible to establish a risk assessment and management basing on data composition and identification of the nano-risks in the life cycle of nano-based products (EC 2004c). Therefore one of the most important fields in the present nano-research is the nanotoxicology with immuno-, pneumo-, nephro-, neuro- and cytotoxicity. And one of the main nanotoxicological research fields is the nanopollution, i.e. the distribution of nano-products and nanoparticles in the environment, the exposure of living organisms and humans at the work places. With nanotoxicology and nanopollution is connected the discussion about the scope of implementation of the precautionary principle.

## 5. Nano-Safety and Nanotoxicology

Nanotechnology as "the platform for 21$^{st}$ century technologies" refers to the most important issues in ethical and societal dimensions such as the health effects and environmental impacts, and the efforts "to use nanotechnology to clean our environment and improve public health", which in context of public communication are pivotal with regard to the possibilities of nano-research commercialization, because: "The public's enthusiasm for an emerging new technology can easily turn to fear, with grave consequences for commercialization. Emerging technologies do pose risks that are ill-characterized, and the best thing nanoscientists can do – both for the discipline and society – is draw attention to possible risks and study them carefully" (Colvin 2003a). The main challenge hereby is to prove the toxicity of nanoparticles and nano-based product. One of the first research on toxicity of engineered nanoparticles were initiated at the US Rice University's Center for Biological and Environmental Nanotechnology in 2001 with the aim to develop of exposure guidelines as the first element of quantitative risk assessment, the second element shall consist of elaboration of models encircling the toxicology of the major classes of nanomaterials. The nanotoxicological question concerns the biodegradation and long-term persistence in the environment of the nanomaterials, then the interactivity of nanomaterials in living systems, e.g. by entering into organism. The lack of technical and empirical data creates hereby a situation of uncertainty, so that "it would transform the business and research enterprise of nanotechnology much as it did those of agriculture

## C. Ethical Aspects: Nano-Safety and Nanotoxicology

biotechnology" (Colvin 2003b: 1166). In this manner research on nanotoxicology appears as condition of establishment the nano-industry from the one hand, and from the other hand nanotoxicology makes out the decisive part of assessment of the environmental and health impacts of engineered nano-based materials and products. What underlines Colvin, nanotechnology "has a unique opportunity in the history of technology: this could be the first platform technology that introduces a culture of social sensitivity and environmental awareness early in the lifecycle of technology development" (Colvin 2003a).

Nanotoxicology as a new subdiscipline of toxicology is established in the last ten years with focusing of research on the question of the safety and the potential toxicity of nanoparticles (NP) and nanomaterials (NMP), including the analysis of particles toxicity and its impact on workplace, environment and consumer safety. Nanotechnology as a new emerging and enabling technology contained also a new kind of hazard, risk and potential harm. In this context was proposed a new subcategory of toxicology – nanotoxicology "defined to address gaps in knowledge and to specifically address the special problems likely to be caused by nanoparticles (…) that protocols should be developed for testing of all materials in the nanoscale, where there is the potential for substantial human exposure" (Donaldson et al. 2004: 728). The first protocols are elaborated for the inhalation testing of particles. But there still remains a lack of measure of the different potential toxicity of nanoparticles, what results from the unfinished research on the nanoparticles properties at all, e.g. the relation of size and surface area on the deposition. The risk with use of nanoparticles results from the greater potential or ability to translocate through the organism comparing with the other materials and particles, for instance nanoparticles entering the blood or the central nervous system can affect cardiac or cerebral functions. These risks underline the necessity to constitute a new subcategory of toxicology and a new subdiscipline of nanotechnology. With this thought in mind the aim is to develop nanotoxicology as a support and completion of a safe and sustainable nanotechnology industry.

Nanotechnology has diverse impacts on materials, engineering, environmental and informational technology, health and pharmacy. But from the other side this new technology needs a complex assessment regarding to the risks: "Any technology before introducing it to the marketplace and into the product chain needs careful evaluation with regard to its sustainability and risk perception" (Donaldson et al. 2004: 727; cf. one of the first assessment of nanotechnology by Paschen et al. 2003). By the way, this assessment shall include also the social dimension of the new technologies and products, because there is a gap of knowledge and reflection concerning the new technologies (nano-bio-info) as the subject of humanities and social sciences. The engineering and life sciences and its basic research are hereby the forerunner of development comparing to the humanities, which have to react to the new results of research and technological innovations. This applies especially to the ethical concerns of nanotechnology: "As the science leaps ahead, the ethics lags behind. There is danger of derailing NT if the study of ethical, legal, and social implications does not catch up with the speed of scientific development" (Mnyusiwalla et al. 2003: R9).

From begin of the nanotechnology development in the 1990s still remains the question how "to map the risks and opportunities" appearing with them. By the risks it is a matter of the potential hazards arising from: "While some of the products that will contain nanoparticles are likely to have them fundamentally bound up in the structure, there is the potential for exposure to NP and nanomaterials throughout the product chain during manufacture, application, and waste management; subsequently there is a need for a toxicology that can assess the likely harm they may cause" (Donaldson et al. 2004: 727). At the same time the particle toxicology is a mature science and has as objective "mechanisms of lung injury caused by nanoparticles"; but there is no size cut-off particles' harm, because "harmful particles have their effects as a consequence of two factors that act together to determine their potential to cause harm: their large surface area, and the reactivity or intrinsic toxicity of the surface": "It is self-evident that the smaller particles are, the more surface area they have per unit mass; therefore any intrinsic toxicity of the particle surface will be emphasized" (Donaldson et al. 2004: 727). As examples of the toxicological hazard of nanoparticles there are cosmetics and sunblock cream with dermal exposure, nanoparticles in food and their ability to cross into the gut lymphatic, nanoparticles using by diagnostic and therapy of drugs. This concerns especially carbon nanotubes with unusual toxicological properties which can cause lungs harm, i.e. "the same material in the form of NP is more toxic than in the form of larger, still respirable, particles": "Very small particles are smaller than some molecules and could act like haptens to modify protein structures, either altering their function or rendering them antigenic, raising the potential for autoimmune effects" (Donaldson et al. 2004: 728). – But from the other side if smaller means really to be more toxic? "The hypothesis that smaller means more reactive and thus more toxic cannot be substantiated by the published data. In this respect nanomaterials are similar to normal substances in that some may be toxic and some may not. As there is not yet a generally applicable paradigm for nanomaterial hazard identification, a case by case approach for the risk assessment of nanomaterials is recommended" (EC/SCENIHR 2009: 9).

Besides nanotoxicology in the last few years is established a second new discipline in the nano-domain: neuro-nanotoxicology with the question if nanoparticles induce neurodegenerative diseases, i.e. the understanding of the origin of reactive oxidative species and protein aggregation and misfielding phenomena in the presence of nanoparticles. The main issues hereby are: 1) passing the barrier between blood and brain; 2) oxidative stress in living system caused by nanoparticles (effects: cell and DNA damage) and protein fibrillation with the end-effect of neuro-degeneration; 3) how far engineered nanoparticles "present a significant neuro-toxicological risk to humans"; 4) development of "a simple screening and risk assessment matrix for nanoparticles in neurodegenerative diseases" for instance in the case of Alzheimer and Parkinson; 5) constitution of a new discipline in the nanoscience: neuro-nanotoxicology (cf. Riediker and Katalagarianakis, 2010: 182). The constitution of neuro-nanotoxicology results also from the development of the neuro-science and above all the new subdiscipline – the neuro-ethics – in the last decade.

But with nanotoxicology and nanopollution the main problem of research remains "still unclear exactly how nanomaterials interact with biological entities and which parameters of the nanomaterials drive these responses" from the one hand, and the "exposure of workers and consumers to nanomaterials" from the other hand (Riediker and Katalagarianakis 2010: 100). This is also the main challenge for risk assessment and management. Henceforth it is necessary to elaborate standard protocols and tests for the assessment of potential hazard by using of nanomaterials, what means to integrate risk into the life cycle assessment of the products based of nanomaterials, but also methods of measurement by nanoparticles exposure.

Dissemination of using in industrial and consumer products of the engineered nanoparticles (ENP) poses the question concerning the health effects, the potential health risks for workers and consumers. The aim of the different projects is an elaboration of risk assessment of engineered nanoparticles and its implementation. The risk assessment of engineered nanoparticles can base on the traditional risk assessment adapted to the exposure-dose-response paradigm for ENP by inhalation, ingestion or dermal exposure in the target organ. In this manner it is a new risk assessment approach corresponding to the specific and uncertain analysis of ENP. Therefore the risk assessment of ENP consists of: 1) Hazard identification with a comprehensive set of measurement and development of "multidisciplinary sets of tests and indicators for toxicological profiling of nanoparticles (NPs) as well as unravelling the correlation between the physicochemical characteristics of NPs and their toxic potential on various organs of the human body" with assessment of risk by industrial manufactured NPs (CellNanoTox-project; Riediker and Katalagarianakis 2010: 3–4). 2) Dose-response assessment with testing systems (in-vitro or in-vivo); and especially by air pollution "creating and validating new instruments and assays to assess the possible toxicity of nanoparticles and to detect nanopollution at occupational sites, in order to promote safe NP manufacturing and handling", and then "detection of NP effects on human cells" (DIPNA-project; Riediker and Katalagarianakis 2010: 12). 3) Exposure assessment, i.e. implementation of ENP with environmental impact especially in aquatic system.

At the same time in the present state of knowledge still remain major gaps regarding to the health and environment impacts and risks resulting from the use of nanomaterials. This concerns the exposure of workers during the production, recycling or disposal process, then of consumers by using the nano-based products, and finally the environmental risks of nanomaterials. Therefore two main aspects with regard to the toxicity of nanoparticles and nano-based products are considered: 1) "to study the evaluation of nanomaterials (physical and chemical properties) during their life cycle and their toxicological impact on human health and environment related to this evolution", then 2) "the development of innovative strategies for the disposal of nanocomposites at the end of their life cycles" (Riediker and Katalagarianakis 2010: 136).

The projects concerning safety of nano-research are characterised by similar results with synergistic outcome. Above all it is the better understanding of the impact of nanoparticles on health and the environment, what enables the definition of future actions in the fields of research and politics. The aim is "to improve the understanding of the potential environmental/health impacts of nanotechnology-based products over their life cycle", and "gathering and generating data on the possible impact on human health and/or the environmental impact derived from the use, re-use, recycling and/or final treatment and disposal of nanotechnology-based products containing engineered nanoparticles" (Riediker and Katalagarianakis 2010: 136). All the projects concerning the safe development and use of nanotechnologies have the aim "to contribute new knowledge to what will be a global endeavour in addressing the scientific uncertainties related to the health and environmental effects of engineered nanoparticles and to provide a body of new information and a new tool that industry and governments can use to begin to assess the risks of these nanomaterials" (Riediker and Katalagarianakis 2010: 142; EC 2005). The aim of the started hitherto different projects is the establishment of a universal framework of risk assessment and risk communication developed adequately to the specificity of nanomaterials. The other leitmotif of the different research projects is to design the first devices for monitoring the safety on work places, environment and the air in public buildings with automatic evaluation of engineered nanoparticles toxicity.

The risk assessment concerns the unique properties (e.g. surface reactivity) of engineered nanoparticles with numerous new applications from the one, and the lack of knowledge concerning their physical and chemical properties, and the levels of exposure from the other hand. In this context the challenge is to elaborate a method and device of ENP monitoring "to separate ubiquitous background nanoparticles from different sources from the ENP" (Riediker and Katalagarianakis 2010: 74). This shall make it possible to differ between physic-chemical and toxicological properties of the ENPs, and to indicate their potential toxicity and bioactivity. The aim is to develop a method of characterising ENP at work places with a portable and easy-to-use device, to develop "methods for calibration and testing of the novel devices in real and simulated exposure situations" (Riediker and Katalagarianakis 2010: 74), also as condition of the safe use of engineered nanoparticles and nano-based products.

In the case of nano-risk-assessment and nano-impacts on health and environment the most important role play the patterns of social communication in form of an "intuitive toxicology", i.e. "how an inexpert or lay audience comprehends and reacts differently to expert information, in this case quantitative toxicology data" (Allhoff and Lin 2009: 91). In consequence the public communication about nano-risk can include totally different and incomparable elements of assessment regime, "that the public uses a non-rational calculus based on a matrix of attitudes and beliefs" and the experts "do not include these non-rational variables" (Allhoff and Lin 2009: 92). The public often act not irrationally but non-rationally confronted with the problem of human or ecological nano-toxicity. The problem consists of elaboration of a communication model encircling researchers, the lay public and the intermediaries such as media, politicians, non-governmental organizations, interest-group and industry. Each of these communication target groups can refer to totally different background of information where motifs and metaphors play a pivotal role.

The term "intuitive toxicology" was introduced in the 1990s by the comparison of public and expert judgments concerning chemicals and risks involved with. In generally, the intuitive toxicology "refers to the assignment of risk which involves biases that may exclude both probabilities and assessments of hazards quantified by empirical research" (Allhoff and Lin 2009: 94), what underlines the differences in perception of risk: the perception by public is more intuitive and less objective, but at the same time multidimensional, i.e. it contains values and prejudices, ideological point of view, and individual experiences; the risk-perception by experts is or shall be purely basing on quantitative data as condition of rationality, in consequence it has to be unidimensional and basing only on empirical data[38]. The other distinction concerns the danger perceived as real, and the risk as socially constructed. In consequence, there can appear totally different modes of interpretation and analyses of risk, that the "data and information from experts and the media are decoded by the public using an algorithm that was not used by the experts when encoding the information" (Allhoff and Lin 2009: 96). At the same time public risk perception is connected with perception of benefits resulting from a new technology or products, so that "it might be possible to change perceptions of risk by changing perceptions of benefits" (Alhakami and Slovic 1994: 1096; quoted by Allhoff and Lin 2009: 96). This result from the key mediators in risk perception by public opinion that besides familiarity, equity, knowledge and responsibility of future generation (sustainability), there appears the next one: the voluntariness of exposure. This attitude depends on the totally different hermeneutics avenues confronting with technological convergence at all. By the experts it is the dosage and exposure, and by the public the consequences and implications. Especially by perception of new risk the trust and distrust are important by elaboration of a communication strategy. But both trust and distrust are above all social construction, what applies also to the case of 'certainty and uncertainty'. Finally, the risk by nano-risk communication is the negative stigmatization of nano-domain as a whole. Hereby appears the ambivalence but also importance of mass media communication.

## 6. Ethical, legal and societal implications of nanotechnology

Concerning the ethical aspects of nanotechnology it should be underlined that "there are hardly any genuinely new ethical aspects raised", but there are a set of new challenges such as: 1) the scope of impact of the precautionary principle; 2) the research focuses on the interface between biological and technical systems; 3) the concept of converging technologies with the endeavour to enhance the human performance. The ethical questions concerning nanotechnology are above all interdisciplinary focused, but the "proposal of an independent 'nanoethics', however, seems exaggerated" (Brune et al. 2006: 15). This statement results from nanotechnology understanding as "a new category of technology" with "the precise manipulation of materials at the molecular level", or on scale of 1 to 100 nm, exploiting "novel properties that emerge at that scale", whit the unique phenomenon "that ordinary materials can have extraordinary properties", what makes it possible that "by precisely manipulating common elements at the nanoscale, scientists can fashion new materials" so for instance the carbon nanotubes (Allhoff

---

38  Cf. to this presuppositions the critical arguments presented by Bińczyk, pp. 42–48.

and Lin 2009: xxiii). But it is also the question how far nanotechnology is a distinct discipline in the system of science? "[T]here is still a debate over whether nanotechnology is an independent or new science, so unique from other fields that it should require or deserve its own category or moniker. Some have complained that nanotechnology is not distinct from other sciences – or at least its boundaries might be somewhat hazy – and therefore its ethics must be equally ill-defined. Others argue further that nanoethics is not an interesting or distinct field because it does not raise any new questions that are not already considered by, say, bioethics or computer ethics" (Allhoff and Lin 2009: xxiv). The fact is that nanotechnology is "a convergence or amalgamation of several existing disciplines" (Allhoff and Lin 2009: xxv). In this manner nanoethics "seem to raise new ethical issues insofar as it adds a new dimension" (Allhoff and Lin 2009: xxviii) to the existing hitherto problems, e.g. the privacy, equity and the new possibilities of surveillance. From the other side nanotechnology still remains an emerging technology visionary understanding as molecular manufacturing including a kind of print-on-demand products to a nano-factory, i.e. "if we can precisely manipulate molecules, and physical objects are only made up of molecules, then why wouldn't we be able create any physical object we want?" (Allhoff and Lin 2009: xxix). But this futuristic vision expresses also the prematurity of nanoethics itself.

One of the most important, and at the same time the new subject with respect to the social condition of technological development and ethical reflection on nanotechnology are the presented until now 'nano-visions', so that there appears a new kind of assessment – visions' assessment – so for instance by the vision of crossing the borderline between humans and technological artefacts or systems, especially in the case of biotechnology and information technology. The other question appears by the medial communication about nanotechnology and the risks involved. Besides 'nano-visions' as the integrated element of nano-assessment and public opinion appears finally the question how to educate about nano-domain. With the social and medial communication about nano-risk is connected the problem of technological education, especially in the field of nanoscience and nanotechnology – with nanotechnology as an objective of learning and teaching. First of all it is necessary to develop an interdisciplinary reorientation of the curricula focused on physics of small systems or nanostructured materials as background for an interdisciplinary openness. Nano-visions, public opinion and nano-education make out the background of the possible recommendation by shaping a socially robust nano-domain as science and technology. The question is how far the scientific factor plays hereby the decisive role. The last element under considerations is the scope and strategies of nano-commercialisation. The commercial needs in the field of nanotechnology include above all the issues concerning patent analyses and higher education and research policy (R&D programs), where the questions are focused on: 1) patents law and practice; 2) professional intellectual property management as pivotal for the successful commercialisation; 3) shaping an intellectual property regime and intellectual property infrastructure at research institutions with envisaged and guaranteed effective patent protection as one of the main tasks of the policy framework. Hereby the patent practice appears as condition of protection and commercialisation in the field of nanotechnology[39].

Following to this a complex analysis in form of nanotechnology assessment shall be widespread, i.e. basing on dialogue and knowledge exchange across the borders of scientific disciplines and the fields of society and culture. In this point of view the public opinion about technologies such as nuclear energy or genetics is crucial: "Neither must risks be denied, nor must chances be overestimated – otherwise much harm could be done. An honest and transparent discussion is the basis for widespread acceptance of any new technologies" (Brune et al. 2006: 9). This concerns especially the case of nanotechnology with disappearing borderlines between established disciplines in the modern system of science and knowledge. Hereby the starting point of an assessment are the questions, how far properties and functions of materials at the nanoscale are changing in comparison to the ongoing hitherto research. In consequence the question is how far these properties and functions can be used for technological applications, but also commercialisation, above all concerning the information storage and (nano)biomedicine.

The philosophical and humanistic reflection about nanotechnology concerns ethical, legal and societal implications with the question how far a specific nano-ethics is needed (Mnyusiwalla et al. 2003; Khushf 2004). The interests on nanotechnology in society resulted from the accelerated development, but from the ethical point of view there are not serious objections till now. Moreover the presented nano-ethics approaches are focused on the visions instead of the real results and the ongoing research. Following to this there is a need for complex nano-assessment, where ethical issues make out an important part, but this shall result from the ongoing research and not from presented futuristic visions of nanotechnology itself. Nevertheless, the visions play a decisive part with respect to public perception and opinion about this technology. Scientific development

---

39 At the same time we can observe in the last decade a kind of patents' pathology on global measure, cf. OECD *Scoreboard* from 2011.

## C. Ethical Aspects: Nano-Safety and Nanotoxicology 153

of nanotechnology poses ethical questions above all with regard to the conception of converging technologies, where nano-ethics is completed by bio-ethics and neuro-ethics.

But as for the autonomy of nanoethics Allhoff argues "that nanoethics lacks any metaphysical autonomy (from other areas of applied ethics), but I nevertheless think that the field can receive a pragmatic justification. I take this pragmatic justification to be weaker than a metaphysical one, but a justification nonetheless" (Allhoff and Lin 2009: 4). Nanoethics consists of research of the ethical and social dimension of nanotechnology, and nanotechnology is the ability to understand and control of the appearing phenomena at the nanoscale, so the argument by Allhoff, "where unique phenomena enable novel applications", but this novelty results not only from operating with 'small things', but from "properties that are manifest because of the nanoscale" (Allhoff and Lin 2009: 5). From the ethical point of view in the nano-domain appears the problem of nano-equity, i.e. fair distribution of new technologies and their use in global comparison concerning environmental protection, medical application, and safe and clean energy, where the decisive principles results from the distributive justice. Nano-surveillance is the next relevant objective of ethical and social impacts of nanotechnology, and one of the most important fields of nanotechnology applications with regard to the ethical and societal impacts – the nano-medicine with three main applications: treatment, diagnostic and drug delivery. But in generally the nano-ethical issues "are not tremendously novel, though they will have be addressed within a new context" (Allhoff and Lin 2009: 33), so that a kind of pragmatically oriented and justified nano-ethics can contributed to the better nano-perception by the public opinion.

Ethical and societal approach to the nanotechnology includes following steps of considerations: 1) the indication of criteria of ethical relevance as for technological development, 2) the application of these criteria in the field of nanotechnology with mapping out the ethical challenges resulting from development, use and mass dissemination of nano-based products and treatments. Generally in the field of nanotechnology the ethical issues are not completely new and have been discussed in other contexts. The new topics appear with regard to the improving human performance and the general human enhancement in the field of converging technologies and are involved in the considerations devoted to the technological convergence at all (Roco and Bainbridge 2003). Ethical relevance appears hereby with the assessment of the presented visions and is focused on the category of responsibility and responsible dealing with these new technologies. Moreover, the ethical reflection on technologies at all concerns first of all the fields and manners of applications and the observed and foreseen impacts on humans, society and environment. And this is a challenge in the case of ongoing development of research and technologies, where the reflection is confronted with a gap of knowledge, where it is impossible to mark rationally the side effects and unforeseeable negative aspects of a technology.

The background of ethical reflection on nanotechnology is the distinction between factual morals and ethics "as the reflective discipline in cases of moral conflicts or ambiguities" (Brune et al. 2006: 401). This distinction underlines the significance of the meta-ethics understanding as critical reflection, which results from the plurality of morals as constitutive element of the modern societies, and expresses the relations between science, technology and ethics, where ethics contributes to management of conflicts resulting from different moral attitudes. This distinction between factual morals and ethics and therewith introduction into analysis of the societal impacts of technology become significant in the case of the ethics of technology in a pluralistic society, where the normative aspects of science and technology are the objective of the social debates. But which are morally relevant contents in science and engineering, and which should be the ethics of science, engineering and technology? With the rejection of the value-neutral conception of technology in the 1990s began the discussion about the possible ethical framework of science and engineering, especially concerning the normative framework of technology development including laboratory research. At first, science and technology are recognised as the integral part of societal processes, embedded in social and political reality: "In this sense, there is no 'pure' technology as a technology completely independent of this societal dimension. Technology, therefore, is morally relevant concerning its purposes and goals, the measures and instruments used, and the side-effects evolving. Currently, technology is recognised as an appropriate subject to moral responsibility" (Brune et al. 2006: 402; Rapp 1999), and the socially robust and post-normal science can be denoted also as responsible one.

It is crucial for the ethical reflection about technology which types of decision-making situations are recognised as a 'standard' one, and which criteria can be applied for classifying them. These criteria should define above all the framework and limits of such standard situations. And the conditions of standard situation with moral and ethical respect can be formulated as following: "steps, decisions and processes in shaping technology are free from the demand or necessity for ethical reflection if, and only if, they take place in a 'standard situation' with respect to moral issues. A standard situation is given if there is a comprehensive, clear, consistent, commonly accepted and factually followed normative framework, which has to be and factually is followed in the specific context" (Brune et al. 2006: 403). The criteria of a standard situation are hereby:

1) comprehensiveness with norms, principles and customs making out the normative framework; 2) clearness with rejection of ambivalence by understanding of the normative framework and at the same time with clear assignment of responsibility; 3) local consistency without contradictions as a base of acting; 4) the general acceptance of the normative framework as base of technology acceptance, which is developed accordingly to this framework, but which is an object of discussion and time-limited accordingly to the technological changes; 5) compliance factually acknowledged with application of and following in practice accordingly to the normative framework in laws, regulations and moral codes by persons and groups. This normative framework is a set of axiological information for all, who are shaping and involved in technology, and this consists of: 1) political regulation with environmental and safety standards and law regulation by designing and development of new technologies, and 2) quasi-regulation by customs and traditions in form of ethics codes. In the case of nanotechnology the ethically relevant aspects arise at the moment, where moral conflicts or ambiguities impair the existing framework of standard situations in moral aspects.

To the main issues with regard to the ethical relevance of nanotechnology belong the problems of privacy, man-machine relation, and equity. But generally there is not a complex analysis, including multiple research streams on nanotechnology, focused on ethical and societal implications: "An independent ethical perception of nanotechnology is currently nearly not recognizable. (…) The lack of dialogue between research institutes, granting bodies and the public on the implications and directions of NT may have devastating consequences, including public fear and rejection of NT without adequate study of its ethical and social implications" (Mnyusiwalla et al. 2003: 11). In this manner the starting point by exploration the ethical dimension of nanotechnology is the question, which are the genuine, new and specific ethical topics in nanotechnology? Ethical reflection on nanotechnology is more general, and analysis of concrete application concerns the nanoparticles, their risks and chances, and recently the possibilities of their medical application.

This applies to the manufactured nanoparticles which are not present naturally in the environment such as fullerenes and nanotubes. There is relevant also the distinction between nano-ethical dimension and nanotoxicology as a scientific subdiscipline of toxicology: "Questions of toxicity for the environment and for humans (…) are, however, not ethical questions. In these cases, the pertinent scientific disciplines, such as toxicology or environmental chemistry, are competent". Following to this the analysis of possible risk of nanoparticles "is not an ethically relevant question per se", and as a standard situation in the case of risk are the established risk regulation measures: "Ethics comes into the game as soon as this will be not the case, so that existing risk regulation may be insufficient or inadequate. The main question then is about the applicability of the precautionary principle and its consequences for practical action" (Brune et al. 2006: 407). Hereby the normative and ethically relevant aspects of application of precautionary principle in the case of nanotechnology are related with: 1) the political dimension with regard to the criteria for action and decision-making; 2) the epistemic dimension with the uncertainty of the scientific knowledge; and 3) the societal dimension with the degree and scope of acceptance and the criteria of social acceptability of technological development and the side effects.

Following to these the question is if the lack of knowledge justifies a kind of moratorium concerning nanoparticles as product of mass consumption accordingly to the principle of responsibility, because the application of the precautionary principle shall result from expectation founded by knowledge, what includes the criteria of concrete benefits and hypothetical hazards. How far the comparison of possible risks of nanoparticles is justified with other technological risks, especially in comparison with the natural nanoparticles such as volcano eruption, and the exposure to manufactured nanoparticles such as fullerenes, nanotubes, smoking or barbecue, and how far these manufactured nanoparticles are the new type of risk as in the case of asbestos? (Ball 2003). In this manner ethical reflection on nanotechnology, and especially on nanoparticles, focuses on a value judgement appearing in a situation, i.e. which values are involved in, which are possible moral conflicts, and how far the precautionary principle is to be used. The main research field and objective of ethical reflection about nanotechnology are the questions concerning "the acceptability and comparability of risks, the advisability of weighing up risks against opportunities, and the rationality of action under uncertainty" (Brune et al. 2006: 409). This is the new field of application in the ethics of science and technology with an interdisciplinary oriented research including toxicology and legal regulation with risk assessment.

The development of nano-domain extrapolated ethical challenges which express also the specificity of nanotechnology such as the equity and the sustainable development from the macro social scale, and from the individual point of view the controversies with respect to the privacy issues and the new possibilities of surveillance. In the case of equity the ethical question focus on the distributive justice with the transfer of both the risks as well the benefits, and in the case of the sustainability principle the question is how far nanotechnology contributes to the societal inequality as well as in the inter-generational and intra-generational discussions and conflicts. Inter-generational dimension of nanotechnology concerns above all the technological impacts on environment and its contribution to the protection of natural environment with regard to the life-conditions of the coming generation in the future, so that ethical recommendation are integrated in the science and research policy and R&D programs on the supra-national level. Hereby the decisive part of nanotechnology assessment in respect to the principle of sustainability is the life cycle of a technology and its products (Grunwald 2002). But in the case of nanotechnology it is still a sustainability 'potentials', but what is important, the technology itself does not guarantee these, the principle of sustainability should be taken into account with regard to the framework programs of research and development. Intra-generational dimension concerns the distributive justice, because each technology and technological progress itself increases existing inequalities of goods distribution, so for instance in the case of the access to the new nanotechnology-based medical therapies and treatments. Following to this the question is how far nanotechnology will contribute to a 'nano-divide' in analogy to the 'digital-divide' in social life. The discussion on distributive justice become in this context a new relevance on the international level with tensions between technologically advanced and developing countries. In this manner the problems of equity belong to the ethical reflection about nanotechnology.

One of the new topics regarding to the ethical dimension of nanotechnology is the problem of privacy, i.e. the new possibilities of monitoring and control technologies developed in the field of nanoelectronics (miniaturization of devices) with new capabilities of data transfer and storage, and new possibilities of citizens surveillance. This miniaturization concerns the nano devices on the point of junction between converging technologies and their military use (cf. Altmann 2004): "But what is not speculation is that with the advent of nanotechnology invasions of privacy and unjustified control of others will increase", and this result from the point of view of technological determinism: "when new technology provides us with new tools to investigate and control others we will use them" (Brune et al. 2006: 411; Moor and Weckert 2004). The problem of privacy is expressed also by the new medical devices (the lab-on-a-chip) and the ability to foreseen and transfer personal health data. The possibilities, for instance, of decoding the complete individual genetic dispositions, can become an economic factor in the medical services, job chances and might be relevant for insurance companies. The question of privacy is connected with a new dimension of societal equity and equality from the one hand, and the new appearing menaces (e.g. the new form of terrorism) and data protection problem on the other hand.

One of the most important research and application fields in nanotechnology, which includes directly the ethical issues, is medicine. Nanotechnology based treatments in the field of medicine make its possible the earlier discovery of diseases or predispositions to these with the diagnosis basing on the 'lab-on-a-chip' as a form of personalized medicine. These new possibilities are expressed by the targeted treatments, nanoparticle dosage system, biocompatibility of implants or drug delivery systems and minimisation of side effects. With respect to the ethical and toxicological dimensions the use of nanotechnology in medicine is recognised as a 'standard situation', but there appear new types of responsibility, so for instance by new forms of drug delivery via fullerenes, nanostructured membranes, gold nanoshells, dendrimers accordingly to the Health Technology Assessment (HTA) till now. But in generally medical applications of nanotechnology do not encircle specific ethical concern. At the same time nanotechnology, biotechnology and neurophysiology and their nano-based synthesis function as background of the posthuman vision of nanotechnology development.

With these new possibilities in the nano-based medical treatments and with the advance of the technological convergence, so for instance in the case of nanobiotechnology, the borders between technology and life, between living organisms and technological devices, are blurred: "Basic life processes take place on a nano-scale, because life's essential building-blocks (such as proteins, for instance) have precisely this size", and supported by nanobiotechnology "biological processes are made nanotechnologically controllable. Molecular 'factories' (mitochondria) and 'transport systems', which play an essential role in cellular metabolism, can be models for controllable bio-nanomachines". In this manner with nanotechnology are given possibilities of a new type of 'engineering' on the cells-levels, so that an "intermeshing of natural biological processes with technical processes seems to be conceivable" (Brune et al. 2006: 415). Following to this nanotechnology seems to be an interface between the mechanics and the basic living, natural and organic units.

This dynamics is the driving force of research in the field of new materials production and in the medicine, and the diverse opportunities involved in refer to the new philosophical, ethical and societal aspects of nanotechnology. The safety standards and risk assessment depend however on the possible applications and products. One of the most important ethical issues is the various possibilities of misuse with the new generation of biological weapons based on technically modified viruses. The main field of research concerns the possibility of crossing between the technical and natural boundaries, between technical systems and human nervous system with the aim to connect molecular biology with nano-engineered devices, so for instance by compensation of damage by sensory organs and nervous system or by increase the capacity and spectrum of human perception. The development of nano-informatics with manufacturing of implants enters into the capabilities of natural systems of life. Following to this a new definition of prevention and misuses is needed.

The development and progress in the field of nanotechnology in the last ten years extrapolate the following issues: 1) the ongoing convergence of technology and humanity, which became much more subtle and conceivable; 2) the philosophical question concerning the necessity of change of the self-conception of humanity and the philosophical conception of man confronted with the realistic assumption of possibilities to create for instance an artificial brain in the next 25 years (cf. Roco and Bainbridge 2003: 256–260). Consequently, one of the main streams of the philosophical reflection on nanotechnology consists of the various philosophies of life and the conception – more or less rational – of transhumanism. An ethical reflection 'in advance' on this subject is presented by the conception of converging technologies focused on the interfaces between nano-bio-info technologies and cognitive sciences, where the 'brain-research' appears as crucial for technological development, and where the philosophical, ethical and societal issues are at first formulated. The background of this assumption is the ongoing debate about artificial intelligence and life.

Nowadays it is the question of human enhancement as the main motive of the concept of converging technologies, which changes at first the understanding of technology itself as only a means of augmentation of human capacities and possibilities. In this sense converging technologies overcome the existing hitherto distinction between natural and artificial beings, where the technological factor became a decisive part of the possibilities of development and human enhancement instead of education or upbringing. It is not only technologically supported medical 'healing' or 'repairing', but 'remodelling' and 'improvement' of human body, where the extension of the physical and sensor capabilities can determine the mental constitution of man. From philosophical point of view the fundamental distinction between 'healing' and 'enhancing' appears with the necessity to redefine conception of 'health' and 'illness' (Habermas 2001). This is no more a 'standard situation' as in the case of nanoparticles, the potential included in the converging technologies radically changes the understanding of man, society and culture. The last step of these consideration concerns the nanotechnological possibilities of overcoming aging and the postponement of death as the leitmotiv of visions in nano-domain (Moor and Weckert 2004: 307–308). This point of view (nano-optimistic visions) is negated by the traditional philosophical ethics (Habermas 2001; Hook 2004). Nevertheless, the human enhancement and improvement of human performance are technically supported and created in the field of nanotechnology, which appears in this sense as an enabling technology with new constitutive ethical questions (Khushf 2004).

Therefore is connected the question if a nano-ethics is needed? And what is the objective of nanotechnology? It is "the ability to technically manipulate matter at the level of atoms and molecules", and at "this level of abstraction such an activity seems to be morally irrelevant: atoms and molecules cannot claim special moral rights", and do not have a moral status comparing for instance with the moral status of stem cells and the passing debate of ethically founded limits of research and applications. Besides this argumentation nano-domain as a research field disperses in various nanotechnologies which could have different side effects of application and therefore different ethical issues and moral aspects, so that "there seems to be no longer a focus for a coherent ethically motivated objection of the entire nanotechnology" (Brune et al. 2006: 421, 422).

## 7. Nanoethics or explorative philosophy of nanotechnology

In the presented hitherto ethical analyses of nanotechnology many authors state that there are not genuine new issues. At the same time the analyses concerns the meanwhile established speculative nanoethics and its legitimacy, i.e. the ethical analyses focus on visions and scenarios of nano-development instead of the really ongoing nano-research and the possibilities of a nanotechnological convergence. In this way appeared the dilemma – nanoethics: needed or superfluous. In this context A. Grunwald postulates instead of nanoethics the development of an explorative philosophy of nanotechnology. In favour of constitution of a new subdiscipline of ethics – the nanoethics – argue for instance Moor and Weckert (2004), Khushf (2004) and Allhoff and Lin (2009), in opposition are Grunwald (2010), Fleischer (2011), MacDonald (2004), and Baumgartner (2004): "We don't need any new subdiscipline of applied ethics called 'nano-ethics' but there is a need for ethics in and for nanotechnology because some topics and questions gain larger urgency and will converge in nanotechnology, and because nanotechnology accelerates scientific and technical progress" (Brune et al. 2006: 433). The nano-domain is characterised by an intrinsic heterogeneity in particular to the wide range of possible applications, so that it is needed ethical consideration of nanotechnologies in certain fields but not a new field of 'nano-ethics', and this ethical considerations can participate and contribute to the process of technology design in the future, so that the two complementary orientations by technology assessment 'ethics at last' and 'ethics at first' can be integrated with the regard to the impacts of the ethical reflection on the technological development. Above all the relevance of ethical reflection and impacts in advance to the technological developments is underlined in the model of 'ethics at first'. In this manner ethical reflection shall be an integral part on the basic-research level. From critical point of view is underlined that the model 'ethics at first' is too speculative and abstract from the ongoing research and possible applications. The postulated alternative is, "that instead of speculative nanoethics we should better speak of and develop explorative philosophy of nanotechnology" (Grunwald 2010: 91).

The speculative orientation at the beginning of nanoethics development extrapolated at the same time the gap between ethical reflection and the technological and scientific progress in nanotechnology in the last decade: "but a new gap has opened up because most nanoethics is too futuristic, focusing on nano-enabled devices that can read our thoughts, for example, at the expense of ongoing incremental developments that are more ethically significant", so that "[i]t is important to consider the ethical aspects of nanotechnology, but it is equally important to ensure that these considerations do not end up as speculative ethics" (Nordmann and Rip 2009: 273). The ethics itself works with the hypothetical cases ('what if' and 'if then'), but this is not adequate approach by the analysis of science-based technology development as for the nano-domain: "As the hypothetical gets displaced by a supposed actual, the imagined future overwhelms the present" (Nordmann and Rip 2009: 273), what happened in the case of public opinion about nanotechnology which are dominated by futuristic scenarios (cf. the "if and then syndrome": Nordmann 2007: 32; Roache 2008). In the case of speculative nanoethics paradoxically is reversed the relation between science and ethics, that as the speculative ethics "leaps ahead" the current science and ongoing research "is left behind, and in this sense "we need to encourage discussions about quality of promises", i.e. a new kind of assessment – the nano-visions' assessment – as an objective of ethics, because "everything that is physically possible is not always technically feasible" (Nordmann and Rip 2009: 274). And in the DEEPEN project we find the postulate: "Move away from speculative debate! Return 'ethical concerns' to the sphere of politics!" (DEEPEN 2009: 7). From the other hand A. Grunwald points out: "If this diagnosis is true, then large parts of nanoethics are misguided and concern themselves with irrelevant and purely speculative ideas, while the really important developments are not taken into consideration" (Grunwald 2010: 92). In his essay on nanoethics Adam Keiper concludes in this context that from the one hand "no scientific field or technological innovation has ever faced such intense scrutiny so prematurely", but from the other hand with regard to the nanoethics it "bears all the signs of prematurity. Its time may come someday, but it is too soon to say just when and how" (Keiper 2007: 55, 67). Nevertheless the question remains, if nanoethics is really too futuristic and speculative? The decisive part of nanoethics deal with particular issues in the current development of nanotechnology: "We should distinguish between philosophical reflections on more speculative questions in nanotechnology ('explorative nanophilosophy') from ethical inquiry of more practical issues ('applied nano-ethics')", and the main task of the explorative nanophilosophy is "to prepare the ground for future debates conceptually, theoretically and methodologically, e.g. by analysing and reflecting alternatives in creating, regulating and using new technologies. (…) Such explorative philosophy must be epistemologically informed instead of being merely speculative" (Grunwald 2010: 92).

The arguments using by critics of speculative nanoethics are among other (Nordmann and Rip 2009): 1) nanoethics is too futuristic, 2) nanoethics is 'oversupply' comparing to the demand of ethical reflection in the field of nanotechnology, 3) there is too much speculative nanoethics, and 4) nanoethics itself is too speculative. The objection that nanoethics is too speculative refers to the implausible nature of the developments, that nanoethics assumes to come from nanotechnology: "nanoethics even goes so far that it apparently takes developments into consideration that contradict the laws of nature" (Nordmann and Rip 2009: 273). Following to this nano-ethics represents hypothetical assumptions: "Indeed, it is not unusual for a discussion of the consequences of nanotechnology to include a second – or third – level conditionality, namely when these consequences might occur as a consequence of the use of nanotechnical products that themselves only might or could become reality, and then only if the respective technical development would take place in the direction envisaged" (Grunwald 2010: 93), so that prospective orientation by analysis and assessment is partly speculative.

The background of the debate focused on ethical aspects of nanotechnology is hereby the principle of responsibility presented by H. Jonas (*Das Prinzip Verantwortung*, 1979) and the strong formulation of the precautionary principle: "If the consequences and side effects of technology could not be evaluated prospectively in any manner whatsoever, the question of responsibility would be superfluous: that which is not known cannot be subjected to any ex ante ethical reflection, and no one can be held responsible for it". Following to this, ethics would be obsolete: "The object of ethics would get lost in the speculativeness of the considerations of consequences, making ethics obsolete: ethics would not set limits on technology, but technology – as a result of its uncertain future – would set limits on ethics" (Grunwald 2010: 93), and this understood as a preventive critics of criticism of speculative nano-ethics before this was appeared (Nordmann 2007). The dilemma consists of two models of ethics towards technological development and accelerated progress, if the 'ethics at first' or 'ethics at last' should be used (Moor and Weckert 2004). These two models shall be completed by a vision-assessment strategies which will have to design the limits of speculative approach in the nano-ethics itself. The two additional arguments presented by Nordmann and Rip (2009) against the speculative nano-ethics are:

## C. Ethical Aspects: Nano-Safety and Nanotoxicology 163

1) the artificial creation of concern as objections to nanotechnology, and 2) the opportunity-cost argument.

Ad 1. Artificial creation of concern as nanotechnological objections. The argument "that stood at the beginning of nanoethics, namely the anxiety that unanswered ethical questions could case nanotechnology to lose acceptance"; i.e. there was a gap between nanotechnological risk assessment and ethical reflection on nanotechnology and the technological and specific progress of nanotechnology itself, what could result of societal, political and public rejection of this technology: "Study of the ethical issues and of the consequences of nanotechnology is thus regarded as being necessary in order to be able to introduce innovation in modern societies. Its absence might otherwise result in the threat of public resistance, which would hinder both progress and the social utilization of the expected advantages and benefits of science and technology" (Grunwald 2010: 94). In consequence: "The only way to avoid such a moratorium (…) is to immediately close the gap between the science and ethics of NT. (…) Either the ethics of NT will catch up or the science will slow down" (Mnyusiwalla et al. 2003: 12); the conclusion is "the earlier that the possible ethical problems from nanotechnology innovations are recognized the more possible it might be to avoid them constructively" (Grunwald 2010: 94).

This is applied more to the model of 'ethics at first', but from the point of view of the model 'ethics at last': "the supposed arbitrariness of speculative nanoethics could lead to completely unfounded and unjustified ethical problems which then could become issues for nanotechnology in the real societal debate" (Grunwald 2010: 94), and being completely artificial creation it could determine the public and political acceptance or rejection and therewith development of this technology at all: "worries about the most futuristic visions of nanotechnology can cast a shadow on all ongoing work in nanoscience and technology" (Nordmann and Rip 2009: 274). In conclusion: "A premature and thus inevitably, at least in part, speculative handling of the ethical questions of nanotechnology would produce artificial problems, whose communication to the public might cause damage to innovation policy in the field of nanotechnology and could prevent exploiting its chances", and that means, nanoethics "should deal with 'realistic' concerns rather than with speculative one" (Grunwald 2010: 94). But the dilemma between 'ethics at first' and 'ethics at last', between 'mere speculation' or 'justified concern' (or plausible future scenarios) remains as a framework of nano-ethics and ethical reflection on technology at all. Nevertheless, "the degree of rationality of an assumed future concern obviously has direct ethical and political implications" (Grunwald 2010: 95).

Ad 2. In the argument of opportunity costs is extrapolated the fact that speculative ethics and visions of nanotechnology "making them unavoidable for other issues" (Nordmann and Rip 2009: 273). This concerns the understanding of nanoethics as applied ethics. From the other side the speculative topics in nanotechnology are integrated above all in the STS studies, what means at the same time a negation of the opportunity costs' argument.

Confronted with the discussion's issues in the last decade the fact is that 'nanoethics' appears as a "misleading umbrella term" (Grunwald 2010: 96), because there are different analyses of the societal impact of nanotechnology, and more of these analyses do not belong to the field of applied ethics. The applied ethics encircles such topics as equity regarding to the benefits of nanotechnology or implementation of precautionary principle (Grunwald 2008b), but also the more 'speculative' visions assessment focusing on the human/technology relationships. Grunwald proposes in this context the term 'explorative nano-philosophy' instead of 'nano-ethics', what makes it possible to integrate all the philosophical, societal and ethical issues connected with nanotechnology development and what can moderate the nano-debates in the future. Especially in the relationship between ethics, science and technology development the ethical reflection often come too late and has no possibilities to shape the development of a technology. In this context in the STS studies is postulated to integrate ethical, philosophical and societal aspects of technology development on the basic research level: "The hope was and still is that addressing more directly the earlier stages of development would allow for a greater contribution to shaping the technology" (Grunwald 2010: 96).

From the other side there appears the so called Collingridge (1980) dilemma "according to which it is in principle easier in the early phases to influence the course of events, but in the early stage of development the required knowledge is absent that would enable one to intervene in a constructive manner"; in this context Nordmann and Rip (2009) diagnosis expresses, "that nanoethics, in order to avoid the Scylla of the Collingride dilemma (namely of coming too late), must fall victim to the Charybdis (i.e. inevitably being too speculative)", but this dilemma (first or last, earlier or too late, realistic or prospective/visionary/utopian) is not an alternative of either-or, "but a differentiation of ethical reflection in line with the problem at hand and with the validity of the available knowledge of the consequences"; and above all the ethical reflection serves different purposes: "The initial question is which purposes ethical reflection on nanotechnology is supposed to serve. Only then would it be possible to determine whether and under which conditions 'speculative nanoethics' might help to satisfy at least some of these purposes or to find that it is obsolete" (Grunwald 2010: 96).

The earlier and even speculative reflection concerns conceptual, hermeneutic, epistemological, anthropological and philosophical issues, where the ethics makes out a small part. "Ultimately this leads to the recommendation not to speak of 'speculative nanoethics' but to consider these forms of reflection as elements of an explorative philosophy of nanotechnology with an own justification, with own rationales and objectives, and with a different methodology". This explorative philosophy of nanotechnology is also the first level by approaching to the new phenomenons and issues resulting from the development of a new technology, but this first and earlier speculative evaluation of new development in science and technology "is in no way supposed to provide orientation for action in areas of concrete developments": "Its task is rather to prepare for possibly coming debates, in a conceptual, theoretical and methodological sense as well as with a view to basic distinctions and relationships" (Grunwald 2010: 97) such as man-technology, life-technology or nature-technology. Grunwald proposes following classification of the fields in the explorative philosophy of nanotechnology.

I. Nano-Epistemology with the "expectations of a new unity of science" with integration of physics, chemistry, biology and the engineering sciences into the nanoscience accordingly to the postulate to shape the world atom by atom: "Epistemological reasoning questioned the validity of this simple analogy and the 'atomic reductionism' behind it" (Grunwald 2010: 98).

II. Nano-Anthropology and the relation between human and technology concerning the new field of nano-biotechnology with life processes on the nanoscale with proteins and DNA combining biological modules on the molecular level and producing functional building-block including technical materials, interfaces, and bounding surfaces – in analogy to 'shaping the world atom by atom'. Hereby cells and their organelles are interpreted as micro- or nanomachines (as part of mechanical engineering), what means a nanotechnological infiltration of molecular biology, genetics and neurophysiology: "The nanotechnical (possibly feasible) duplication of fundamental life processes is the essential prerequisite for crossing the borderline between technical and living systems", and this has a basic importance for new interpretations concerning the convergence of humanity and technology with the question of borders between humans and technology "with respect to their transmissibility". At the same time there appears a need of new kind of self-understanding and self-concept of humanity, "whether and to which extent this would increasingly summarize human beings under the realm of technology" (Grunwald 2010: 98; cf. Grunwald and Julliard 2007).

III. Nanotech-Hermeneutics with philosophical interpretations of nanotechnology including the changes of human culture and the relationship to the nature which could be technologically based and mediated: 1) new Baconism with the attitude of 'shaping the world atom by atom' with optimistic message "in making everything technically possible", and with the *homo faber* figure who become able "to control everything based on his or her control at the level of atoms and molecules. Humans would become 'engineers of evolutionary processes' and divine makers of the world" (Grunwald 2010: 98). 2) Nanotechnology understanding as crucial factor of the new type of manufacturing with radical uncertainty of societal impacts and risks: "Everything could be possible – and probably nothing could be controlled" (Grunwald 2010: 99). 3) Futuristic vision with nanotechnology interpretation as a "cipher of the future": "In this way, explorative philosophy can prepare the ground work for applied ethics and for the technology assessment of the developments when they become more concrete. Ultimately this promotes a democratic debate on scientific-technical progress by investigating alternative approaches to the future of humans and society with or without different nanotech developments" (Grunwald 2010: 99).

IV. Epistemology of explorative philosophy of nanotechnology: 1) methods of visions and promises assessment with respect to rationality: "It needs clarification of the cognitive and normative content of the partially speculative future conditions and consequences of nanotechnology"; 2) with the main task of "an epistemological 'deconstruction' of these future conditions in order to be able to qualify the object of a subsequent ethical reflection with regard to its applicability and validity"; 3) and finally visions' assessment: "building block of an open, cognitively informed and normatively oriented dialogue, such as between experts and the public or between nanotechnology, ethics, research funding and regulation" (Grunwald 2010: 99; cf. Roache 2008).

## D. Convergence and Nanotechnology Assessment

### 8. Converging technologies: Categories, principles and fields

The mentioned at the beginning the huge perplexity of emerging techno-sciences is increased in the field of technological convergence. And the research on converging technologies takes place not only incidentally in programs and projects which are interdisciplinary oriented. On the contrary, the changes in the system of science in the last two decades are global in measure and express the development of techno-sciences at all. The driving force of this process is even the technological convergence which has changed the whole configuration of the relationship between science, technology and society (Bainbridge and Roco 2007). And the decisive part of this development takes place in biotechnology and biomedicine involved in the research at the nanoscale. Following to this the progress in the nano-domain (nanoscience and nanotechnology) can be seen as condition and driving force of the converging development, for instance the case of new ICT generation and nanoelectronics with the started in the last few years research on graphene, in nanotechnology and biotechnology it is the improvement of information processing capabilities and the 'construction' of molecular nanomachines (Wang 2013), in the materials science there is the emerging additive manufacturing technology (Gibson et al. 2010), and finally cognitive sciences basing on nanotechnology and focusing on the efforts to inquire the functions of neurons at the nanoscale (e.g. the EU-Flagship research project on human brain). These projects are the 'second generation' of research on the global scale comparing to the GMO and the human genome projects in the 1990s. All of them are founded or established by the scientific expansion of the research on the technological convergence.

At the same time the ongoing research on technological and scientific convergence has impact on all fields of culture and social life, so that the changes of science and technology influence the condition and development of society itself. The convergence can be understand at the first approach as the process of discovering and establishing of interrelations and interfaces between nano- and biotechnology, information technology and cognitive science, which open the new fields of applications with regard to human enhancement (Bainbridge and Roco 2006; Müller 2010), including the ethical, societal and political dilemmas and challenges. In this manner as one of the most important issue appears the question concerning risk or/and impact assessment and management, which supposes changes in the way of thinking about technological development and methods of its analyses. For instance the linear model of innovation and the static distinction between basic and applied sciences is replaced through systemic, interactive and holistic approaches, which set the framework of technology assessment, technoscience and the concept of converging technologies.

The other issue concerns the role of citizens by risk assessment and management, i.e. which shall be the participative framework of civil and democratic society with respect to the science and research policy and the development's strategies in these pivotal fields. Especially from political point of view the emerging importance of implementation of the precautionary principle is crucial in the relationship between science, technology and society. The precautionary principle basing on scientific evidence has hereby two functions, it makes out the possibilities of risk assessment, but also the development and assessment of alternative solutions with regard to the safe use confronted with possible harmful side effects. Moreover the precautionary principle includes two aspects, the societal and then the political one as a category of research policy and R&D programs. Therewith the main issues confronted with the development of converging technologies are human enhancement and intellectual property rights from the one hand, and the precautionary principle as the main reference regarding to the participative attitude of citizens confronted with the expansion of technological convergence as the background and condition of establishing an active "scientific citizenship" (Mali 2009: 54; Bucchi 2004).

From the point of view of technological convergence the tasks of precautionary principle are: 1) indication of uncertainties appearing at the interfaces between society (culture) and nature; 2) indication of uncertainties and unforeseeable side effects of human enhancement and 'invasive technologization' of human life including the use of everyday life technical gadgetries (Böhme 2008), and the biomedicine and biogenetics basing on the nanoscale treatments and therapy with the dilemma of distinction between health therapy and unjustified human enhancement (Habermas 2001). Hereby as the dominant critics appears a new kind 'bio-luddism' and techno-conservatism with underlined negative aspects of transgression from 'therapy' to 'enhancement' with a new (possible) form of discrimination consisting of the genotype (Fukuyama 2002: 123; Hughes 2006: 288). And the new form of discrimination may concern the health care services, life insurance policy, labour law and market, reproductive right and family law and policy. Therewith the new type of technology-based medical treatments and the potentiality of technological human enhancement are relevant for the policy-making process on international level (cf. UNESCO 2003).

In the perspective of risk management technological convergence appears as a concentration of different uncertainties, and could be regarded as a "model uncertainty" which "is inherent in scientific systems with multiple variables interacting in complex ways" (Mali 2009: 57). In this context the precautionary principle can be understand as: 1) implication of "a broad range of possible measures calibrated to the degree of uncertainty and the seriousness of the consequences that are feared" (Whiteside 2006: 53); 2) that decision-makers in the field of politics are able to act in advance of scientific certainty, and establish of a preventive attitude to protect society from harm in the case of accelerated progress; 3) and with distinction between 'precautionary' and 'preventive' policy: "prevention consists of action taken to reduce known risks, while precaution aims to anticipate and reduce more uncertain risks" (Mali 2009: 58). In the process of policy making this means to cross from the attitude of 'reaction' to the use of precautionary strategies. Moreover the preventive attitude is characterized by the model of 'ethics at last', and the precautionary principle is more associated with the model of 'ethics at first'.

Following to this there are two main tasks of precautionary principle, the establishment of interfaces between separated disciplines and their researchers, and then the introduction of transparency between research and public opinion.

These means to fulfil two criteria: 1) Precautionary principle is a policy approach focused on cross-disciplinary problem-solving which results from risks involved in the present science and research, and this presupposes an elaboration of a new transdisciplinary conception of risk as a base of cooperation between different disciplines and their theoretical and methodological background and specificity. This transdisciplinary cooperation appears as condition of the converging technologies and their social impacts analyses. 2) Precautionary principle expresses also the ongoing debate concerning public participation in the research policy, also as a decisive factor of democracy: the precautionary principle endeavours "to do this by translating the divergent preferences of various stake holders involved in S&T matters into effective policy choices", what underlines the changes in the R&D policy decision-making: "The shift from 'government' to 'governance' in science and technology presumes the active mobilization of the public in science risk assessment and science risk management. The concept of 'governance' reflects the currently changing modes of political action. By talking about governance, political actors are no longer referring to top-down decision-making. Instead, they are talking about interdependence, networks, and partnerships" (Mali 2009: 59).

By precautionary principle "priority is given to clarifying the gaps in knowledge and identifying early warning signs and unintended consequences of actions", and the "scientific evidence is used not only for the diagnosis of risks but also to develop and assess safer alternatives to potentially harmful consequences of scientific activities" (Mali 2009: 61). Following to these statements the precautionary principle was misunderstood and finally rejected by the US R&D policy. In opposition, in the EU it makes out the research policy (EC 2000), especially in the field of converging technologies (cf. HLEG-report from 2004 where the precautionary principle makes out the strategy of EU and the *European Knowledge Society*), that the great scientific opportunities can be identified with various types of risks.

The HLEG-report differentiated between various types of risk resulting from the development of converging technologies: "1) the risk of investing in technological promise that does not materialize; 2) the risk that consumer acceptance of new technologies outpaces the careful consideration of their consequences; and 3) the risk inherited by converging technologies through the contribution from various enabling technologies" (HLEG 2004: 39). In the European R&D policy is established therewith the model of democratic participation concerning science and technology development. In US this remains still an objective of scientific, military and political elites, what means also a technocratic and paternalistic model of R&D policy. This 'antagonism' expresses the distinction between EU and US policy-making processes in the field of science and research. It seems that the European political culture is based more on the principle of responsibility. The US research policy as market and application oriented (with commercialisation, commodification, patent law and practice) created the post-academic system of science and research basing on an enlarged cooperation with industry, where the traditional scientific professional ethos such as presented by Merton (1973) with the norms of community, universalism, disinterestedness, and organised skepticism is marginalised (Ziman 2000). Following to these changes is presented the postulate to elaborate a scientific citizenship or the proactive concept of citizenship (Bucchi 2004) understanding as the right of citizens "to test or to contest expert claims and even to produce alternative options" (Mali 2009: 72). Bucchi (2004: 108) postulated a replacement of the deficit diffusionist model of the traditional relationship and communication between science and the public by a dialogue between citizens and experts. This is also the new mode of shaping the relationship between science, knowledge and politics.

The concept of scientific citizenship results from the increased societal implications involved in the development of technological convergence. Hereby the analyses of societal implications in the case of nanotechnology encircle: 1) Examination of trends and opportunities appearing in nanoscience and nanotechnology with regard to the societal impacts, benefits and risks, and 2) summing up of the newest realm of science and technology focused on nanotechnological development. The first such complex examinations were presented by the workshops at the US National Science Foundation (NSF) since 2000 (Roco and Bainbridge 2001), and were focused on 10 main topics: 1) productivity and equity, 2) future economic scenarios, 3) the quality of life, 4) future social scenarios, 5) converging technologies, 6) national security and space exploration, 7) ethics, governance, risk and uncertainty, 8) public policy, legal and international aspects, 9) interaction with the public, 10) education and human development (cf. Roco and Bainbridge 2005). The aim of the followed examinations was to present "clear, scientifically grounded statements of the principles of nanotechnology to enable non-scientists to participate effectively in public policy discussion" with underlined significance of nano-education as a field of general technological education for the non-scientists: "Participants were especially concerned about the need for educational reforms not only to teach principles of nanoscience and nanotechnology to students at many levels, but also to support cross-disciplinary training, to equip underutilized scientists and engineers with nanotechnology-related skills, and to train social scientists and other scholars to conduct science and technology studies related to the field" (Roco and Bainbridge 2007: 1, 2). In consequence there remain seven main tasks, topics and categories by the analyses of the nano-domain as science and technology:

1) Economic impacts include the scope of commercialisation of nanotechnology. This concerns above all the growing nanotechnology-based industry: "Pressing challenges include the need to identify the best innovation models for management, the processes that will provide essential human resources and the most informative measures of success" (Roco and Bainbridge 2007: 2). The question is how far there is established a specific nano-economy, especially in the fields of a new generation of nano-based materials and electronics, and how far nanotechnology contributes to the sustainability of economic growth and changes the structure of professions on the labour market.

2) Social scenarios as an anticipation of the nano-domain development including the alternative paths of research and implication of nanotechnology. This anticipation of development in the future is the background of distinguished ethical and legal issues concerning the impacts of nanotechnology such as the problems of privacy and control, and mode of interaction resulting from the development of mobile information technologies: "The question often will be not how a given nanotechnology will affect otherwise stable conditions, but how it will interact with the chaotic forces that swirl in an already unstable world, such as the impending population collapse in postindustrial nations caused by insufficient fertility" (Roco and Bainbridge 2007: 2).

3) Converging technologies. Nanotechnology is recognized as the decisive factor by the development of technological convergence, i.e. creation and establishment of reciprocal relationship with biotechnology, information technologies and their interfaces to the new technologies based on cognitive science with the aim and goal to improve human performance (Roco and Bainbridge 2003). The scope of applications at the nanoscale encircles "molecular machines comparable to the natural machinery inside living cells, medical devices and materials that might be implanted inside the human body, and the application of principles from computerized natural language processing to genomics and proteomics" (Roco and Bainbridge 2007: 3). The question remains how far these are futuristic visions of development, but which need an anticipative analysis of risk and elaboration of possible regulations by law: "The complexity of such multidisciplinary innovation will make it difficult to distinguish real risk from phantom risks in practical contexts such as product liability law". The question concerns the ability of technological convergence which is in opposition to the historically founded system of science: "Convergence of historically distinct sciences and branches of engineering is not an automatic process, and it may be difficult to achieve because the communities involved are culturally heterogeneous

and may unnecessarily defend their professional autonomy. At the same time, convergence of technical disciplines at the nanoscale will be essential for society, for example, by providing the knowledge and the tools for environmental protection" (Roco and Bainbridge 2007: 3).

4) Ethics and law or ethical and legal issues of nanotechnology development are focusing on elaboration of an ethical and legal framework including the development of nanotechnology itself but also its undesirable side effects. The ethical issues in the nano-research concern among other topics the question of transparency in public communication and decisions-making confronted with the political practice on the point of junction and cooperation between research centers and industry and the volume of research public funding. The legal aspects of nanotechnology concern the regulation by law with regard to the dissemination and mass consumption of the nano-based products and the protection of intellectual property rights and patent law in the field of rapidly changing nano-research.

5) Governance of nanotechnology and the necessity of appropriate mechanisms development in the field of politics, because political regulations encircle both the possibilities and paths of applications and dissemination, then the public funding of technology, science and research at all. The political dimension of nano-domain expresses the main objective of the multidisciplinary field of science and technology studies, which are focusing on research of the interfaces and relationship between the constitutive domains of the modern society confronted with the phenomenon of post-academic science from the one hand, and the technological convergence from the other hand.

6) Public perception of science and technology, especially concerning the newest development in the field of nanoscience and nanotechnology, is characterized by the participative turn and the model of deliberative democracy, where the public opinion and perception of technology become a decisive factor in the policy-making process. "However, public involvement cannot be a one-way street, in which our only responsibility is to make sure that correct information flows to the public. Rather, the public must be fully engaged, with ample opportunity to express its concerns and state its goals for the technology" (Roco and Bainbridge 2007: 4).

7) Educational issues in the field of nanotechnology include the curriculum development and multidisciplinary framework elaboration, especially by engineers' higher education. The fact of the technological convergence underlines the necessity of a new framework of education, "that many students must be trained to be interactional experts mediating between disciplines", but this educational framework shall be reciprocal: "Many non-technical advanced students

will need to learn about nanotechnology, and future nanoengineers will need to study the ethical and societal implications of their work" (Roco and Bainbridge 2007: 4).

The social impacts of nanotechnology development can be analyzed in two perspectives: 1) concerning the contribution to the economic growth, and 2) with a view to the process of research commercialisation. The promise of nanotechnology is the economic growth and increased productivity as expression of a new age of post-scarcity. It is an analogy to the ITC-revolution (cf. *Magna Carta for the Knowledge Age*, 1994), but this revolution changed a little in the wealth of nations, and the GDP of the developed countries is falling down in the last 20 years: "So there are those who raise the specter of a digital divide spiraling into a further nanotechnology divide. For many advanced industrial nations, nanotechnology is seen as a means of assuring or re-capturing a lead role in the global economy" (Roco and Bainbridge 2007: 114). In this manner nanotechnology will accentuate these divisions concerning social stratification and inequality from local to global level. The controversy concerns above all the visions, promises and critics of converging technologies. But this techno-optimism or enthusiasm has been often critiqued, and the history itself shows us the double-face of technology and partially science. So that nanotechnology is a background for both utopian and dystopian scenarios. The 'realistic' point of view underlines that the scarcity will remain a part of human condition, and technologies caused transformations cannot be identified with prosperity at all. The other phenomenon is the ongoing process of research and science commercialisation since the 1980s (Krimsky 2003; Florida 1999), "that the pendulum has swung too far and that society will play a serious cost if commercialization continues to progress without challenge" (Roco and Bainbridge 2007: 118).

## 9. Principle of technological convergence

In the following are mentioned projects and programs focusing on the assessment and implementation strategies of nanotechnologies in Europe and USA. The starting point is the final report of the EU project CONTECS (Andler et al. 2008) dedicated the multidisciplinary concept of converging technologies and in particular with the understanding of nanotechnology as an enabling technology and as condition of technological convergence, i.e. as fundament of the synergy between nano-bio-info technologies and cognitive science from the one, and with analysis of the strategic role of the humanities and social sciences in the development of key technologies from the other hand. The project examined the impact of the converging technologies (Nano, Bio, Info and Cogno, NBIC) on the humanities and social sciences. The main aim was to analyse the role and significance of the humanities and social sciences concerning the new emerging and at the same converging technologies from the one, and from the other hand to underline the phenomenon of technological convergence which directly influence the present cognitive sciences. The main subject was to formulate the ethical, societal and political implications of these converging technologies, but also to extrapolate the key opportunities and challenges with them in the future the modern society will be confronted. That means also an initiation of a permanent monitoring of the development in the field of converging technologies, also as an orientation for research and development activities and policy on national and international level. In this sense it is an attempt to characterise the point of junction between the humanities and social sciences from the one hand, and engineering sciences and its converging technologies from the other hand. This concerns above all nanotechnology as the fundament of the phenomenon of technological convergence.

The concept of technological convergence or converging technologies is formulated in the last few years. It underlines the importance of relations, synergies or fusions between the new emerging technologies and the sciences. The notion convergence means "the coming together of scientific disciplines to solve problems common to these disciplines, initially through interdisciplinary co-operation". Therewith 'convergence' denotes "the development of technologies drawing on a combination of research findings from different disciplines", and it matters especially the transdisciplinary research and development projects "which are situated at the intersections of the key areas of science and technology" (Andler et al. 2008: 1). The base of the concept of technological convergence is the model of NBIC (nano, bio, info technologies and cognitive science) elaborated 2001 in US (cf. Roco and Bainbridge 2003). In the NBIC tetrahedron are expressed interrelations between these key technologies and sciences at once with the convergence between them understanding as a "synergistic combination of four major provinces of science and technology" (Andler et al. 2008: 1), and this combination, the overlapping and the reciprocal interdependence between them, is a condition of its development, which bases on the principle of interchangeability on the nanoscale of atoms, circuits, DNA code, neurons and bits (cf. Bouchard 2003: 12; concerning humanities and social sciences in NBIC-model cf. Bainbridge and Roco 2006).

In this manner the discussion about converging technologies can be understood as a "forum for exploring the future impact of all science and engineering" (Andler et al. 2008: I). The discussion concerning the converging technologies was started in US 2001 with research programs and policy emphasising the central issue of human enhancement, i.e. the possibilities of technological augmentation of human capabilities, then technological modification of human corporeality, finally with the open question concerning technological influence of human mental condition. In the same time European Commission constituted a High-Level Expert Group *Foresighting the New Technology Wave*. In the following time is formulated the CTEKS-agenda (*Converging Technologies for the European Knowledge Society*; cf. the report *Converging Technologies – Shaping the Future of European Science*, HLEG 2004; NEST 2004; the other initiatives cf. Rader et al. 2007; TAB 2008).

In comparison the US-initiative is strongly focused on the individual dimension of human enhancement, and the European programs are shifted more on the societal and environmental needs adequate to the model of knowledge-based society with key position and role of the converging technologies as a condition of a sustainable development (cf. HLEG 2004). From the methodological point of view the European starting point is characterised by an interdisciplinary orientation, whereas the US-initiative reduces the concept of converging technologies to the nano-level interpretations of human enhancement, also as an expression and example of a new unity of science at all. The US-approach in the field of technological convergence transformed finally to a technologically determined societal and political program basing on the universal principle of convergence, that the technological convergence shall moderated also the social and political process of social changes and appears as the strategic element by the process of becoming a society (cf. Roco et al. 2013). In both cases (the US and EU approaches) is emphasised the necessity of identification the research fields which are founded by technological convergence and interdisciplinarity. There appears also the question about contribution of humanities to the converging technologies and "toward understanding the processes of scientific and technological convergence, to analysing the new modes of knowledge production, to assessing the diffusion of the concept into research programmes and to grasping the role of visions in concrete technology development processes" (Andler et al. 2008: 3–4). From the other side the question is how far converging technologies influence and change the humanities and social sciences by remaining the principal challenge for the science at all.

Generally the present discussion on converging technologies is characterised by three main aspects: 1) the concept of convergence inside the system of science focusing on particular issues; 2) the concept of interdisciplinarity with analysis of science and society interrelations; 3) the ethics with the human enhancement issues. In this manner the major aim of the CONTECS-project was elaboration of such an agenda in form of distinguished areas of overlapping or point of junction between humanities, social sciences and converging technologies (nano-bio-info), especially the relevance of nanotechnology, with formulation of the key research questions and challenges: 1) the concept of convergence as an integral part of research and development policies, i.e. as a criterion for financial support of research programmes; 2) the key role of cognitive science and nanotechnology as condition of the development of converging technologies; 3) the relevance of inter- and transdisciplinary orientation concerning converging technologies between the 'hard' and 'social' sciences, because with "the absence of a common basis for interdisciplinary research, it is necessary to study the 'epistemic cultures' of the disciplines involved, in order to identify possible obstacles for cooperation caused by different styles of scientific reasoning" (Andler et al. 2008: IV); 4) technology assessment concerning ethical and societal issues of the converging technologies, especially its application in human enhancement; 5) the problem of naturalised social sciences with a technology-based deterministic conception of man, and oriented again the postulate of preservation of the scientific character of humanities, but at the same time with social science using empirical methods from the natural sciences and vice versa, with the possibilities of a reciprocal completion between the humanities and engineering sciences; 6) the human enhancement as the fundamental issue in the concept of converging technologies and nanotechnology with the definition of the research and implementation targets, but basing on realistic, scientific reasons without futuristic precipitated conclusions.

The general theoretical topic is the techno-scientific process of convergence in the present science and research including the practical issues, which is concentrated on elaboration of proposals for science policy action on the European level for instance in the 8th Framework Program Horizon 2020: 1) Emphasising the strategic role of the humanities and social sciences in the development of key technologies, i.e. integration of the humanities and social sciences in research on converging technologies by financial and institutional supporting; 2) Monitoring of converging technologies development and extrapolating its societal, ethical, political (work and market) significance; 3) Empirical research on techno-scientific convergence as condition and source of decision-making process in the research policy (development of R&D programs, indication of strategic fields and key research); 4) Distinction between disciplinary, inter- and transdisciplinary subjects, but also its overlapping as an expression of the passing reconfigurations in the system of science; 5) Recognition the humanities and social sciences as condition of the enabling and converging technologies development in the meaning of extrapolation the relevant issues, i.e. "identification of critical factors for success under various conditions, and the provision of funding for strongly interdisciplinary and transdisciplinary pilot projects in which SSH [social sciences and humanities] are involved in different ways", and this as guarantee for a) "exploration of new avenues for strongly interdisciplinary cooperation"; b) "strengthening societal and application aspects in the natural and engineering sciences"; c) establishing the humanities and social sciences also as 'converging sciences' (Andler et al. 2008: VI). – Finally the main task is to expose social sciences and humanities as 'mediators' and 'translators' of converging technologies, or perhaps, 'moderators' of the public discussion on converging technologies.

This is also the objective of technology assessment as a way of general characterization of nano-domain. The starting point are the technical capabilities, scientific state of affairs, the political support of research projects in form of government investments and regulations, but finally the development of nanotechnology "depends upon the ways in which existing social, economic, and cultural trends interact with these factors": "The question is not how nanotechnology may change a stable world. Rather, we should ask how the development of nanotechnology will play into the forces that already swirl in an unstable and often chaotic world" (Roco and Bainbridge 2007: 97). With other words, which are the dominant social processes of today that will transform and determinate the society in the future, and how far the passing technological changes are commensurable with the societal transformations, how far nanotechnology can contribute to these. Therewith it is important to elaborate societal scenarios of nanotechnology development as a kind of anticipation of the future.

Concerning societal and ethical assessment the question concerns the right time to insert the assessment what shall be simultaneously with the "technological momentum" (Pool 1997) as a real phenomenon which needs a realistic assessment from the point of view of social sciences and ethics. This approach in the case of nano-domain is characterized by simultaneous analysis accordingly to the technological development from the one hand, and reciprocity and mutual knowledge exchange between the humanities, social and engineering sciences from the other hand. With regard to this cooperation and co-evolution there are distinguished several starting points of view (Roco and Bainbridge 2007: 78–79) such as: 1) competitiveness and safety; 2) ethical issues based on the concept of precautionary principle (Baird 2000; Tickner at al. 2004); 3) the problem of limited resources with technological development as condition and decisive factor of the national security, geopolitics and the power-balance in the future; 4) the scientific facts confronted with the futuristic visions and speculations, and with considerations who is the winner and who the loser by the nano-development, and who decides about the development and the chosen research paths; 5) the strategic point of view is the political one with funding and pressing the nano-research and development strategies; 6) the societal dimension of technological development, i.e. how far the social change can be identified with the technological one (Baird et al. 2004).

The technological convergence became the pivotal field by development of strategies concerning public investments in research at the end of the 1990s. One of the decisive factors hereby was the management of information for converging technologies and their innovations. Following to these new forms of convergence, there is changed the understanding of innovation, especially the so called revolutionary innovation basing on the traditional categories such as 'better, faster, cheaper', which do not change the world as a whole: "For world-changing innovations enabled by convergence, think of innovations that simultaneously act to enable completely new market/customer linkages as well as obsolete technology or production capacity" (Roco and Bainbridge 2007: 139; such as disruptive innovations, cf. Christensen 1997). Technological convergence encircles both the revolutionary as well architectural innovations, where both are objective of basic as well as applied sciences, and in consequence are relevant for industrial base and venture capital. This combination makes out the coordinates of the future development, i.e. the convergence between them. This schema of convergence encircles: 1) basic and applied research, 2) revolutionary, architectural and disruptive innovations, 3) industrial base, 4) mature venture capital industry, 5) prospective higher education and research policy with balancing and equilibrium between 'mission' oriented science and research commercialisation, finally 6) public investments into nascent technology.

The expression and example of technological convergence and the new understanding of innovation are the ongoing research constituting the interdisciplinary field of research and science – nanobiotechnology. Richard Feynman in his speech *There's Plenty of Room at the Bottom* expressed an intuition: "The biological example of writing information on a small scale has inspired me to think of something that should be possible. (…) Consider the possibility that we too can make a thing very small, which does what we want – that we can manufacture an object that maneuvers at that level!" (Feynman 1960: 29) His intuition has been realized by creating micro-machine devices working with molecules at the single scale: "We are starting to understand how cells work and how to manipulate matter on the scale of single molecules. We have been able to minimize and store huge quantities of data and have access to these data" (Roco and Bainbridge 2007: 131). This was also the starting point of development in the nano-domain, especially concerning the diagnosis and new type of medical treatments basing on precise manipulation of matter. As example can be mentioned the first nano-devices such as laboratories on a chip, the micro-tweezers and the long-standing work on neural microprobes, then the micro-cages, pneumatic manipulators, and DNA microarray injection pins, medical nano-shells by cancer therapy, and the first motor being run by a single molecule and engineered basic motility devices, and basing on the MEMS-technology creation of the first robotic device at the scale of micro-technology. The crucial point on this stage of research and laboratory works was to find the interfaces to larger scales as a condition of application, especially with the envisaged next generation type of materials, so in analogy for instance by the reproductive behavior of yeast cells, which is depending on the presence of a single protein (cf. Roco and Bainbridge 2007: 131–132). This is an evidence of emerging functionality at "many nested levels" from supra-molecular systems, cellular systems, organisms to populations. "The implication here is that the living system does not work like clockwork, but comes about as an emergent result of the fact that there is randomness associated with the process. To elicit this higher order of behavior requires us to incorporate this robustness and the randomness associated with it" (Roco and Bainbridge 2007: 132).

In this manner there appeared a paradigm shift with its crucial part – the integrative technology or technological convergence between biotechnology, nanotechnology and informatics: "We look at nanotechnology for the precision engineering of the matter. We look at biotechnology providing the fundamental building blocks for our function. And then we look at informatics for understanding and controlling how the flow goes, between those entities" (Roco and Bainbridge 2007: 132). And this convergence can be incorporate through the nanoscale structures:

> "A living cell is not just a bag of chemicals, but a structure of membranes. Information is incorporated in the structure associated with cells, and that structure provides delineations of functionality that occurs inside the cells. So if you create a biomimetic membrane, a membrane that allows you to insert functional building blocks of living systems, you should be able to get higher-order behaviors. (…) All those functional building blocks are there for us to use like Legos, which can be assembled to create artificial organelles performing unit functions. These are chemical factories, unit operations that function as a result of them working together" (Roco and Bainbridge 2007: 133).

In this manner is exemplarily extrapolated the base of nano-bio-info technological convergence: "This is the very beginning of an induction path. If you have a system in which you incorporate these artificial nodes that self-stimulate, you should be able to mimic the same function that a group of cells does, a higher-order type of function" (Roco and Bainbridge 2007: 134). The next step of convergence concerns the neuronal cells and the brains functions, i.e. the transport of information between neurons and a sensory perception integrating both space and time. – But which are social implications of all these capabilities? First of all it is possible to elaborate a new type of DNA-based diagnosis of disease that may change the medical insurance practice. There are given new possibilities of disease prediction and forecasting of one's health by DNA technology. The technological convergence enables a new approach to manufacturing, i.e. broadcast architecture with 'self-replication' of manufactured goods. With that the hybrid engineering technologies could integrate the natural and inorganic systems (biomimetic system and synthetic materials) in form of biofunctional materials.

In the case of nanobiotechnology the challenge as for public opinion is to take into account the different and particular contexts of social communication consisting of designating the target groups, then "identifying, articulating, and communicating issues to a variety of 'publics' with diverse perspectives, backgrounds, interests, and levels of education, and adjusting communication level, style, and content to particular contexts" (Roco and Bainbridge 2007: 141). The model of communication is constructed as an organizational framework, and as a base of public consultations. It is a set of tools using by explanation of societal impacts of nanotechnology to the public opinion. But the condition of public communication is the construction of scientific discourse which bases on elaboration of a set of primary and secondary literature. Accordingly to John Ziman (1978) a scientific discipline can be developed by discourse characterized by searching for 'consensus' by constitution and presenting primary scientific literature, then following secondary literature layer consisting of comments, references and examinations of primary literature assumptions and thesis. In the case of societal impacts of nanotechnology, and above all concerning converging technologies, the development remains at the first stage, at the primary layer of constitution of scientific discourse about these objectives and primary literature devoted to these issues. The constitution of a new scientific and research field depends also on circumstances representing and working by the particular national research policy and the potential of public investments, so for instance by comparison of the EU member states, but also by the dominant science and research 'cultures' by comparison between EU and US research programs with the degree of cross-disciplinary research projects, then the practice and readiness to cooperate between university and industry, and the potential of venture capital investing in emerging technologies and their research.

In the case of nanobiotechnology and its societal impacts the scientific core is focused on DNA research, including a fusion between biology ('wet') and engineered nanomaterials ('dry'): "Nanomaterials are foreign to biological entities, so the term 'nanobiotechnology' includes integrating the wet and dry

## D. Convergence and Nanotechnology Assessment 181

through material development and processes to bridge the two dimensions", and following to this nanobiotechnology can be characterized as "an emerging area of scientific and technical opportunities. Nanobiotechnology applies the tools and processes of nano/microfabrication to build for studying biosystems. Researchers also learn from biology how to create nanoscale devices" (Roco and Bainbridge 2007: 143). Herewith five major objectives can be distinguished: 1) elaboration of analytical and diagnostic tools using by for instances testing drug toxicity; this include microfluidic and nanofluidic technologies, e.g. lab-on-a-chip or mouse-on-a-chip, in the drug development and clinical diagnosis; 2) highly functional and bio-selective surfaces in form of molecular templates, e.g. self-assembled monolayers using by allergen recognition and by chemical characteristics of biological system behavior; 3) construction of devices using by cell isolation, e.g. cancer cells; 4) building of molecular motors as a type of nano-devices able to move and to deliver the drug; 5) development and implementation of nano-markers and contrast agents to target DNA or other specific biological particles. – Example of the ongoing research concern the microfluidic platform technologies or microfluidic devices ('chips') with following possible applications: 1) clinical and non-clinical diagnosis, 2) pathogen and toxin identification, 3) cancer detection, 4) genetic analysis, 5) portable devices (lab-on-a-chip) used by epidemics and disease outbreaks, 6) externalization of testing processes outside the body with using of microfluidic or nanofluidic devices: "Benefits of externalization include the ability to test biological reaction to a limitless variety of pharmaceuticals and chemicals, and virtually eliminate animal testing" (Roco and Bainbridge 2007: 145). The both sort of micro- and nanofluidic devices make out the possibilities and conditions by development of personalized medicine, i.e. "to analyze personal genetic and biological makeup, and interaction with various forms of proteins, vitamins, and pharmaceuticals for the purpose of proactive, personalized intervention therapy to prevent health problems and optimize personal health" (Roco and Bainbridge 2007: 145). Moreover there appear possibilities of military applications with detection of toxic substances in chemical and biological warfare, then the environmental applications with detection of toxic substances in air, soil and water. At the same time micro- and nanofluidic devices pose legal and ethical controversies such as the control of personal data, privacy, security and ownership which are relevant in the case of convergence between biotechnology and information and communication technologies. But above all the promise of a personalized medicine creates social acceptance. Finally there arises the question of the safe use of them: "It is noteworthy while the devices will fill some needs, they will create other issues, such as a need for disposal systems for the biological and toxic material in the devices. This is an important but undiscussed point in environmental impacts and safety concerns" (Roco and Bainbridge 2007: 145).

In context of nanobiotechnology as an emerging field of science the epistemological dimension appears as decisive factor of scientific discourse. First of all it is the question how far all forms of knowledge are equivalent, because of the new emerging fields of science "(…) are inexact and speculative bodies of research findings derived from a variety of mediated forms of inquiry, including experimental, theoretical, and computer simulation and modeling. There is not and cannot be unmediated, direct human experience at the molecular level. The biotechnology concept of 'substantial equivalence', which refers to considering only outcome and not processes in product development, can also be applied to knowledge. This poses a central epistemological, ethical, and practical question" (Roco and Bainbridge 2007: 147).

How far the three forms of knowledge are substantially equivalent, i.e. theoretical, experimental and computer simulations and modelling? This is relevant by "drug development and pharmaceutical approvals, where both economic and health impact stakes are high". Are these three types of knowledge substantially equivalent and at the same time interchangeable? "Is knowledge derived from externalizing processes through microfluidics, nanofluidics, and microscopic growth chambers substantially equivalent to knowledge derived from drug trials on humans?" And finally: "What are acceptable boundaries and circumstances for recycling on different types of knowledge, and who should determine these?" (Roco and Bainbridge 2007: 147). With these epistemological questions are directly connected ethical dilemmas: 1) "Do similar arguments apply to labeling drugs as to labeling genetically altered food?"; 2) "Should drugs be labeled or characterized according to the type of knowledge utilized in development and regulatory approval?"; 3) "Should there be an economic relationship between types of knowledge and the cost of pharmaceuticals?"; 4) "Where is responsibility for adverse events – in regulatory bodies or manufactures?" (Roco and Bainbridge 2007: 147–148). The starting point by the analysis of technological convergence is to extrapolate the scope of interdisciplinarity in the modern research. This is significant by the process of language modeling basing on analogy between language in everyday life and scientific discipline with experts' language, so for instance in the case of nanotechnology – what really means 'the manufacturing' of nanoparticles or 'nanomachine' or biological 'machine'. Herewith appears the convergence of human language technologies with biological chemistry with the aim to map "of biological sequences to the structure, function, and dynamics of proteins and, ultimately, of biological systems" (Roco and Bainbridge 2007: 153).

Hereby the convergence concerns the analogy between language and biology, and then between information technology and biotechnology.

Encoding of the whole genome type has elaborated the complete lists of all protein sequences which define the function carried out by several organisms, in bacteria and humans. The new research field hereby is proteomics "which looks at all the proteins in a cell simultaneously": "The transformation of linguistics through data availability has allowed convergence of linguistics with computer science and information technology", and in analogy to this "transformation of biological chemistry by data availability opens the door to convergence with computer science and information technology", which happened in the past between bioinformatics and computational biology. The background of those convergences and the envisaged in the future between nano-bio-info technologies and cognitive sciences is the fact, that "human language is, just like biological chemistry – a domain of application for statistical approaches" (Roco and Bainbridge 2007: 153–154). The specificity of language analogies and convergence between language and biology lies in the principle for feature selection, "that experts in language technologies are able to augment current interdisciplinary research in bioinformatics, although the set of tools in principle is in common" (Roco and Bainbridge 2007: 155). Linguistic analogy between miniature machines and functional biological system included in the convergence of nano- and biotechnology is the second example. This is also the new type of convergence between established biological chemistry and nanotechnology as a new scientific domain, where cells and organelles or proteins and protein complexes are viewed as 'machines' such as motor (kinesin microtubule motor), switch (light switch rhodopsin) and pump (proton pump bacteriorhodopsin). The other emerging application field is the environmental protection with a synthesis of applications modes between nanobiotechnology and information technologies.

## 10. Multidisciplinary framework of nanotechnology assessment

In the context of ongoing technological convergence the technology assessment can be understand as „a contribution to shaping technology from the societal perspective" as "an early warning" of potential hazards and "early detection" of potentials and chances of a technology. Technology assessment contributes in this manner to the improvement of society, and in the political dimension to the elaboration of "possibilities for deciding about the future path of technology and the embodiment of technology into society" (Schmid et al. 2003: 99). At the same time technology assessment is a learning process with three levels: 1) the cognitive one with information concerning the impacts of a technology;

2) the normative level including conflicts and problems as the results of technological advance but also with "a learning refinement of ethical sensitivity and of values" (Schmid et al. 2003: 103); finally 3) the societal level of assessment as a learning process focused on risk perception and acceptance by public opinion.

Moreover, technology assessment includes the three major types of innovations, besides scientific and technological also the societal innovations resulting from the ongoing changes in the society, which design a new political and societal framework and administrative structure accordingly to the new challenges, so for instance in the case of establishment of agencies responsible for environmental policy and sustainable development strategies. In this way technology assessment "mostly deals with the innovations promised by materials research, not with the inventions concerning materials or processes directly. This is due to the fact that inventions at the scientific level do not have an immediate and direct impact on society. The impacts on society are arising as soon as ideas for innovation are expected or are coming out from the scientific inventions" (Schmid et al. 2003: 106).

In the case of nanotechnology assessment the nano-domain, including science and technology, then basic and applied research, makes out "an interdisciplinary prototype model" (Schmid et al. 2003: 110) for technology assessment itself. First of all nanotechnology assessment is 'in time' accordingly to the scientific and technological advance and mass dissemination of nano-based products in comparison to the other technologies' assessment such as nuclear energy, genetically modified food or stem cell research. "In this way, nanotechnology assessment (NTA) should cooperate very closely with nanotechnology research, in order to help the innovation system to integrate other social values besides those valid at the economic marketplace" (Schmid et al. 2003: 110). The term 'nanotechnology' introduced Norio Taniguchi 1974 as „production technology to get the extra high accuracy and ultrafine dimensions" (quoted by Fleischer 2002: 113). Following to this nanotechnology as a new research field results from the convergence of the separated hitherto natural and engineering sciences by entering into the nanoscale and by using two methods: 1) physics and electronics with 'top-down' and miniaturization of devices, 2) chemistry and biology with 'bottom-up' and with the aim to build of complex molecular units as clusters or transistors. The innovations in the field of the nano-domain refer to the synergy between top-down and bottom-up approaches, and between physics, chemistry and biology. At the same time 'nanotechnology' is an umbrella term in the science and research policy for all initiated research programs and projects at the nanoscale. In this sense 'nanotechnology' is a cross-cut technology ("Nanotechnologie als Querschnittstechnologie", Fleischer 2002: 115), therewith a key and enabling technology. With regard to the nanotechnology assessment Fleischer distinguishes between: 1) nanotechnology as a supporting ("unterstützende") technology by the development of new materials and innovations' processes; 2) nanotechnology as an enabling technology with new products and treatments; 3) nanotechnology as a condition of systemic innovations. Confronted with the state of knowledge at the moment only the first approach can make out the basis of an assessment.

Accordingly to these the following objectives and tasks of nanotechnology assessment can be distinguished: 1) the substitution of the existing hitherto materials, products, treatments and processes with a set of questions: Which materials can be substituted? Which changes will occur in the industrial system and the labour-market? Which will be the environmental balance and impact of the nano-based products? How far it means the change of the end-treatment and life-cycle of nano-products?; 2) the establishment of the research and laboratory infrastructure and scientific communities, including science and research policy, the patterns of cooperation with industry and the general funding trajectories from the one hand, and the scientific and technological advance itself in form of foresight scenarios or forecasting prediction of development in the future; 3) the concerns in the field of public communication; 4) the environmental impacts focusing on materials flows and life cycle assessment of nano-based materials and products; 5) the ethical dimension of nanotechnology assessment above all with the emerging possibilities in the nano-medicine and nano-biotechnology. Following to these the analyses of nano-domain appear as the development accompanying (nano)technology assessment ("Entwicklungs-begleitende Technikfolgenabschätzung"), i.e. the nanotechnology assessment shall be appointed at the earlier stage of the nano-research and nano-development with the possibilities to design the development paths in societal and political dimensions (Fleischer 2002: 122–124). Therewith a complex nanotechnology assessment shall contribute to the identification of the major development trends in research but also application with the anticipation of the potential risks in the future.

The ongoing research on nanotechnology consists of mapping the scope and issues of cross-disciplinary nanotechnology assessment. One of the most important objectives of the present nanotechnologies research is to define the condition concerning the safe use and manufacturing of nanoparticles and nanomaterials. It is an endeavour to elaborate standard procedures and methods of the measurement and 'classification' of nanoparticles and nanomaterials toxicity (RS/RAE 2004: 69–84). In consequence there appears also the question about the possibilities to precede a complex nanotechnologies assessment. In the first case the research on nanotechnologies led meanwhile to the constitution of nanotoxicology (Donaldson et al. 2004: 727–728) basing above all on the potential risk assessment (EC 2009), then to elaboration the first procedures defining the nano-toxicity accordingly to the law regulations and the international standards of research and development programs (cf. OECD *Series on the Safety of Manufactured Nanomaterials*), finally to the attempts generating the first devices detecting potential nano-pollution in organism, air (public or work places), and aquatic environments (cf. projects in the Nano-Safety-Cluster, Riediker and Katalagarianakis 2010).

The presented in the last few years' solutions in nanotoxicology, the dynamic development of nanotechnology, and the new emerging fields of implementation extrapolate the necessity of nanotechnology assessment as a new technology and science at once. First of all, this assessment bases on an objective (e.g. nanotechnology in physics) and an interdisciplinary (e.g. nanotechnology in physics, chemistry, biology or mechanics) analysis of the state of knowledge and the scope of implementations. In the next step the assessment is extended to a multidisciplinary analysis of nanotechnology impacts in the societal (Kjolberg and Wickson 2010), economic (Kaiser et al. 2010) and ethical (Grunwald 2010; Fleischer 2011) dimensions. At the same time the reference-point of the complex nanotechnologies assessment are the concept of converging technologies (Roco and Bainbridge 2005; Andler et al. 2008) from the one, and the theory of technology assessment (Grunwald 2002; 2008a) from the other hand. The both theories are meanwhile declared as the pillars of the EUs research and science policy by reflecting the distinctions between US and EU approaches (European Parliament 2006), expressed also in the assumptions of the 8th Framework Program "Horizon 2020" (EC 2011a: 46–49) with underlined key significance of nanotechnologies development (EC 2011b). Concerning the societal impacts of nanotechnologies one of the main issues is the analysis of the nanotechnology perception in the media-based debate (Haslinger et al. 2012). The aim of the following considerations is to present a preliminary typology of nanotechnologies assessment and indication of the main implementation fields.

The attempt to elaborate a framework of complex nanotechnology assessment as a condition of technological convergence underlines the relevance of precautionary dialogue concerning the scientific and societal impacts. "However, this multidisciplinary blending of NBIC confronts us not only with auspicious enrichment, but also with analogous potential adverse implications and risks. Risk accompanies innovation and change. Pinpointing the impact of latent risks to these miraculous benefits could take years. But, we must begin the precautionary dialogues now to safeguard society from the possibilities of unforeseeable harm in the future" (Miller 2007: 158–159). The postulated precautionary dialogue poses a set of questions. Which are the "phantom risks" in the nano-domain, and which issues appear as crucial by risk management framework? How far legal processes can control risk connected with converging technologies "when it transcends several areas of law?" (Miller 2007: 159). Which standards of proof shall be used concerning certainty and uncertainty as basis for legal regulations by totally different practical contexts of converging technologies? Is it possible to separate politics from science in risk assessment? How shall be the political framework confronted with the potential risk by converging technologies? At the same time: "There is no opting-out of society in this Age of Convergence" (Miller 2007: 160).

Elaboration of a complex model of nanotechnology assessment and therewith of the technological convergence shall be historically founded above all with regard to the development of engineering sciences such as the materials science, but also the cybernetics and information science in the second half of the 20$^{th}$ century. Moreover, concerning the model of nanotechnology assessment several point shall be taken into account: 1) inclusion of public opinion by defining societal impacts of nanotechnology, science and technology at all; 2) in the case of nanotechnology assessment and the concept of converging technologies the main challenge is to select but also to produce interdisciplinary oriented primary and secondary scientific literature, which should include as well as the scientific and technical knowledge and the philosophical and societal concerns on all disciplinary levels concerning theoretical, methodological and practical implications. In this manner it is possible to construct a socially robust and at the same time responsible nanotechnology (Colvin 2003a). This can be applied also to the technological convergence with information technologies and cognitive science.

From the other side it shall be distinguished between the key research questions and challenges appearing with the nanotechnology as condition of technological convergence. First of all it is the relevance of inter- and transdisciplinary orientation concerning converging technologies between the 'hard' and 'social' sciences, because with "the absence of a common basis for interdisciplinary research, it is necessary to study the 'epistemic cultures' of the disciplines involved, in order to identify possible obstacles for cooperation caused by different styles of scientific reasoning" (Andler et al. 2008: IV). Therewith the main issue in nanotechnology assessment are societal and ethical impacts of nanotechnology, especially by application of converging technologies in human enhancement. In this context appears the problem of naturalised social sciences with a technology-based deterministic conception of man, and oriented again the postulate of preservation of the scientific character of the humanities and social sciences, but at the same time with social sciences using empirical methods from the natural sciences and vice versa, with the possibilities of a reciprocal completion between the humanities, social and engineering sciences.

From the point of view of humanities and social sciences the general theoretical topics in the field of technological convergence are focused above all on emphasizing the strategic role of the humanities and social sciences in the development of the key technologies, what means the integration of the humanities and social sciences in research on converging technologies by financial and institutional supporting. This shall result with a monitoring of the converging technologies development and extrapolating its societal, ethical, political and economic significance. The most important way to analyse the nanotechnological impacts are the empirical research on techno-scientific convergence as condition and source of decision-making process in the research policy (development of R&D programmes, indication of strategic fields and key research). As basically is underlined the distinction between disciplinary, inter- and transdisciplinary subjects, but also its overlapping as an expression of the present reconfigurations in the system of science. In consequence, the analysis and assessment of converging technologies contribute to the recognition of the humanities and social sciences as condition of the enabling and converging technologies development in the meaning of extrapolation the relevant issues, i.e. "identification of critical factors for success under various conditions, and the provision of funding for strongly interdisciplinary and transdisciplinary pilot projects in which social sciences and humanities are involved in different ways", and this as guarantee for "exploration of new avenues for strongly interdisciplinary cooperation", but also "strengthening societal and application aspects in the natural and engineering sciences", and finally establishing humanities and social sciences also as 'converging sciences' (Andler et al. 2008: VI). Exposing social sciences and humanities as 'mediators' and 'translators' of the converging technologies, or perhaps, 'moderators' of the public discussion on converging technologies shall serve by "avoiding counterproductive 'hypes' and identifying societal demands and innovative applications which correspond to policy goals" (Andler et al. 2008: VII).

The main dimensions and fields of nanotechnology assessment are: 1) the typology of technology assessment at all, and 2) the methodology of nanotechnology assessment basing on the comparison with other sciences and technologies. The typology of nanotechnology assessment encircles the disciplinary, interdisciplinary and transdisciplinary scopes or levels of analyses. On each level the assessment is oriented towards theoretical, methodological and practical issues. The theoretical framework of assessment makes out the three conceptions constituting the present philosophy of technology: the technology assessment (TA),

## D. Convergence and Nanotechnology Assessment 189

the techno-science with science-technology-studies (STS), and the concept of converging technologies or technological convergence (CT).

The starting point of nanotechnology assessment is the disciplinary scientific and theoretical assessment, i.e. nanotechnology as the field or subdiscipline of physics what expresses also the historical background of constitution the nano-domain beginning with the speech of Richard Feynman. The disciplinary assessment consists of three stages: 1) the theoretical assessment with analysis and synthesis of the state of knowledge and ongoing research in the physics; 2) the methodological assessment with analysis of the methods and possibilities of their modification and improvement, i.e. the development and construction of new apparatus and instruments, then the development of a new research infrastructure as condition of the nano development at all; 3) assessment in practical dimension concerning the architecture but also implementation of basic research adequately to four phases of research process with design of the research plan, the laboratory experiment, 'manufacturing' of 'artefacts' (nanoparticles, nanomaterials, nanodevices), and finally design and/or indication the scope of implementation.

The interdisciplinary assessment of nanotechnology comprises the nanotechnology relevant issues in natural and engineering sciences, i.e. in physics and materials science, chemistry and biology, electronics, mechanics or environmental engineering. From the historical background it is above all physics and materials science where the field of interdisciplinarity of the nano-domain is established, and nowadays there are the ongoing research on the points of junction between physics, chemistry and biology where the nanotechnological convergence is put in the interdisciplinary measure. On the interdisciplinary level of nano-assessment are distinguished three major stages: 1) the theoretical dimension of assessment with the analysis and design of the scope of theoretical interdisciplinarity; 2) the methodological evaluation with underlying the points of junction between the disciplines and their reciprocal penetration and overlapping, then the exploration of new research issues and objectives, and the indication of the degree of reciprocal exploitation of research results; 3) the practical assessment with the aim to indicate the potentiality to generate new quality of 'artefacts' and devices with the same four stages as well as in the disciplinary dimension of assessment, i.e. with design of interdisciplinary research process, the architecture and planning of laboratory experiments, 'manufacturing' of 'artefacts' and devices, and finally with design and indication of implementation strategies and possibilities.

The third type of nanotechnology assessment concerns the multi- or transdisciplinary dimensions, i.e. nanotechnology as subject of the engineering and life science, philosophy and humanities, social sciences, law, politics and economy, and the use of nanotechnology-based treatments in medicine. Hereby the starting point of nano-assessment is the philosophical evaluation establishing the cross-disciplinary dimension of analyses with three main stages: 1) the theoretical dimension bringing nano-domain in context of metaphysics and ontology, epistemology and theories of science, and anthropology; 2) the methodological dimension with the problem of measurement and observation, and the question concerning the design of research and laboratory experiments including the significance of non-humans elements; 3) the practical dimension focuses on nanotechnology as a subject of ethics. – The next dimension of nano-assessment presents the point of view of social sciences. In the case of theoretical assessment from point of view of social sciences it is the question concerning the impact of nanotechnology on the theoretical models of society in analogy to the information and communication technologies as the background of the theoretical model of informational society with networks and flows, i.e. how far societal macroscale meet the nanoscale. The methodological evaluation underlines the similarities between natural and social sciences by exploitation of the empiric research and the statistic method by analysis of the public opinion concerning technologies development, but also by exploitation of the metaphors of networks, flows, scales, and turns in social theory. The practical evaluation underlines the technological development and the 'enabling' and 'emerging' technologies as decisive factors of social changes in the dynamic process of 'becoming of a society' with the elaboration of strategies by solution of the social conflicts resulting from technologies development from the one hand, and with anticipation of the importance of interdisciplinary experiences in the technologies development and the ability to transgress the boundaries of disciplines.

The nanotechnology assessment from point of view of social sciences is founded by the comparative method. The comparison concerns societies, cultures and civilizations from the one hand, but also all the fields of social life such as economy, politics, law and media form the other hand. Meanwhile a set of issues is presented, for instance 1) the impact of nanotechnology on the particular fields of culture such as politics, education, arts, science, medicine; 2) the significance of technologies in the different civilizations in form of comparative analysis; 3) the role of media in the process of shaping public opinion about technological convergence; 4) affirmation of nanotechnology as the decisive factor in politics (higher education and research policy) and international relation with technology-orientated geopolitics; 5) the degree of juridical regulation concerning manufacturing and use of nano-products, and the impact of nanotechnology on economy and trade theories, management and the market itself; 6) the influence of nanotechnology on natural environment protection and the theory of sustainable development.

In general, there can be distinguished the main criteria of nanotechnology assessment and foresight such as: 1) the scientific and theoretical assessment focusing on the contribution of nanotechnology to the knowledge, methodology and practice in particular sciences; 2) the societal assessment with the analyses of impacts of nanotechnology on the everyday life, e.g. hygiene and medicine, education and science, arts and media, ethics and family life; 3) the economic, managerial and trade (commercial) assessment with the degree of innovation (e.g. the continuative or disruptive innovation), profitability of research and products, exploitation and implementation possibilities in industrial mass production (industrialization of technologies), and technological sustainability towards environment. The societal and political assessment of nanotechnology concerns therewith the acceptance or rejection by public opinion, social trust, security and the scope of legal regulation, but also the degree of risk, hazard and harm, and consequences resulting from mass dissemination of the new nano-based products. In general it is the question concerning the contribution of nanotechnology to social alienation or integration and how far the social life is increasingly intermediated by technologies. Finally, technological convergence extrapolates the dilemma between status quo and social changes, between social conflicts and social consensus concerning technology, and the tension between particularism and universalism of the social groups' interests.

The complex nanotechnology assessment presents the nano-domain as the platform of technological convergence and condition of converging technologies development. Nanotechnology appears hereby as fundament of synergy between nano-bio-info technologies and cognitive science. At the same time there is underlined the assumption of the strategic role of the humanities and social sciences in the development of key technologies. In consequence as the main subject appears the formulation of ethical, societal and political implications of these converging technologies, but also to extrapolate the key opportunities and challenges with them the modern society would be confronted in the future. Herewith is underlined the necessity of permanent monitoring of nanotechnology development and permanent modelling or re-modelling of assessment regime in the case of technological convergence.

Moreover, the presented path of nanotechnology assessment seems to be a condition of a scientific, technological and 'societal' convergence at all. In the context of nanotechnology assessment as the background and condition of technological convergence appears the recently presented approach focused on the convergence of knowledge, technology and society with the understanding of convergence as "the escalating and transformative interaction among seemingly distinct scientific disciplines, technologies, communities, and domains of human activity to achieve mutual compatibility, synergism, and integration, and through this process to create added value and branch out into emerging areas to meet shared goals" (Roco et al. 2013: xiii). This understanding of convergence belongs to the pillars of knowledge-based society and economy, where from the scientific and technological point of view the nano-domain (as science and technology) gives the decisive innovations and put forward the process of technological convergence and development at all. The convergence of knowledge, technology and society bases on the initial convergence at the nanoscale. Therewith is designed a general process of convergence understanding also as an anticipation of changes and challenges in the future. This third stage of convergence explicitly refers to the societal dimension of science and technology: "Today, because science and society are already changing so rapidly and irreversibly, the fundamental principle for progress must be convergence, the creative union of sciences, technologies and peoples, focused on mutual benefit" (Roco et al. 2013: xxiii). The designed process of convergence makes out at the same time the framework of science and research policy in US and Europe (e.g. the 8th Framework Programs Horizon 2020), but also a convergence of basic and applied areas of science and technology, what finally shall enable a technology standardization in these emerging fields. At the same time this understanding of convergence represents a societal and political program with all dilemmas and controversies which will design the societal and technological debate in the future.

# References

Alhakami A.S. and Slovic P. 1994. *A psychological study of inverse relationship between perceived risk and perceived benefit.* "Risk Analysis", No14: 1085–1096.

Allhoff F. and Lin P. (Eds.) 2009. *Nanotechnology & Society. Current and Emerging Ethical Issues.* Heidelberg, New York: Springer.

Altmann J. 2004. *Military uses of nanotechnology: perspectives and concerns.* "Security Dialogue", Vol. 35: 61–79.

Amato I. 2003. *Instant manufacturing.* "MIT Technology Review", November.

Andler D., Barthelmé S., Beckert B., Blümel C., Coenen Ch., Fleischer T., Friedewald M., Quendt Ch., Rader M., Simakova E., Woolgar S. 2008. *Converging Technologies and their impact on the Social Sciences and Humanities.* Final report of the CONTECS project. May 2008. Karlsruhe: Frauenhofer Institute for Systems and Innovation Research.

Bachmann-Medick D. 2006. *Cultural Turns. Neuorientierungen in den Kulturwissenschaften*, Rowohlt, Reinbek bei Hamburg.

Bainbridge W.S. and Roco M.C. (Eds.) 2006. *Managing Nano-Bio-Info-Cogno Innovations. Converging Technologies in Society.* Heidelberg, New York: Springer.

Baird D. 2000. *Organic necessity: Thinking about thinking about technology.* "Techné: Journal of the Society for Philosophy and Technology", Vol. 5(1): 1–14.

Baird D., Nordmann A., Schummer J., (eds.) 2004. *Discovering the Nanoscale.* Amsterdam: IOS Press.

Ball P. 2003. *Nanoethics and the purpose of new technologies.* London: Lecture at the Royal Society for Arts; http://www.philipball.co.uk

Barry A., Born G., Weszkalnys G. 2008. *Logics of interdisciplinarity.* "Economy and Society", Vol. 31: 20–29.

Baum R. 2003. *Nanotechnology: Drexler and Smalley make the case for and against 'molecular assemblers'.* "Chemical and Engineering News", Vol. 81(48): 37–42.

Baumgartner C. (2004), *Ethische Aspekte nanotechnologischer Forschung und Entwicklung in der Medizin.* „Aus Politik und Zeitgeschichte", B 23–24: 39–46. Bonn: Bundeszentrale für Politische Bildung, Beilage zur Wochenzeitung „Das Parlament".

Bennett I., Sarewitz D. 2006. *Too little, too late? Research policies on the societal implications of nanotechnology in the United States.* "Science as Culture", Vol. 15(4): 309–325.

Bensaude-Vincent B. 2001. *The construction of a discipline: Materials science in the United States.* "Historical Studies in the Physical and Biological Sciences", Vol. 31, Part 2: 223–248.

Bhushan B. 2004. *Introduction to Nanotechnology.* In: Bushan B. (ed.), *Handbook of Nanotechnology.* Berlin: Springer, 1–6.

Binks, P. 2003. *Questions loom large in nanotech's tiny world.* "The Age", Vol. 21, October; http://www.theage.com.au/articles/2003/10/20/1066631346170.html

Böhme G. 2008. *Invasive Technisierung. Technikphilosophie und Technikkritik.* Kusterdingen: Die Graue Edition.

Bouchard R. 2003. *BioSytemics Synthesis.* Science and Technology Foresight Pilot Project, STFPP Research Report 4, Ottawa; http://fistera.jrc.es/docs/FISTERA%20CaseStudy_Canada.pdf

Brune H., Ernst H., Grunwald A., Grünwald W., Hofmann H., Krug H., Janich P., Mayor M., Rathgeber W., Schmid G., Simon U., Vogel V., Wyrwa D. 2006. *Nanotechnology. Assessment and Perspectives.* Berlin, Heidelberg: Springer.

Bucchi M. 2004. *Science in Society.* London, New York: Routledge.

Bush V. 1960. *Science: The endless frontier: A report to the President on a program for postwar scientific research.* Washington: US Government Printing Office.

Christensen C.M. 1997. *The Innovator's Dilemma: When new technologies cause great firms to fail.* Boston: Harvard Business School Press.

Collingridge D. 1980. *The social control of technology.* London.

Colvin V.L. 2003a. *Responsible Nanotechnology: Looking Beyond the Good News.* "Centre for Biological and Environmental Nanotechnology", University of Rice; www.eurekalert.org

Colvin V.L. 2003b. *The potential environmental impact of engineered nanomaterials.* "Nature Biotechnology", Vol. 21: 1166–1170.

Crichton M. 2002. *Prey.* New York: Harper Collins Publishers.

DEEPEN 2009. *Reconfiguring responsibility. Deepening debate on nanotechnology*; http://www.geography.dur.ac.uk/Projects/Portals/88/Publications/Reconfiguring%20Responsibility%20September%202009.pdf

Donaldson K., Stone V., Kreyling W., Borm P.J.A. 2004. *Nanotoxicology.* "Occupational & Environmental Medicine", Vol. 61: 727–728.

Drexler K.E. 1986. *Engines of Creation. The Coming Era of Nanotechnology.* New York: Anchor Books.

Etzkowitz H. and Leydesdorff L. 2000. *The dynamics of innovation: From national systems and Mode 2 to a triple helix of university-indiustry-government relations.* "Research Policy", Vol. 29: 109–123.

European Commission 2000. *Communication from the Commission on the precautionary principle.* COM (2000) 1 final, Brussels.

European Commission 2004a. *Towards a European Strategy for Nanotechnology.* COM (2004) 338 final, Brussels; ftp://ftp.cordis.europa.eu/pub/nanotechnology/docs/nano_com_en.pdf

European Commission 2004b. *Nanotechnology – Innovation for Tomorrow's World.* Directorate-General for Research, Brussels, Publication EUR 21151 EN.

European Commission 2004c. *Nanotechnologies: a preliminary risk analysis.* Brussels: Risk Assessment Unit.

European Commission 2005. *Nanosciences and Nanotechnologies: An action for Europe 2005–2009.* COM (2005) 243 final, Brussels.

European Commission, SCENIHR 2009. *Risk Assessment of Products of Nanotechnology.* Brussels.

European Commission 2011a. *Regulation of the European Parliament and of the Council establishing Horizon 2020 – The Framework Programme for Research and Innovation (2014–2020).* Brussels, COM (2011) 809 final.

European Commission 2011b. *Successful European Nanotechnology Research. Outstanding science and technology to match the needs of future society.* Brussels: Directorate-General for Research and Innovation.

European Parliament 2006. *Technology Assessment on Converging Technologies.* Brussels: Policy Department Economic and Scientific Policy/European Technology Assessment Group.

Feynman R.P. 1960. *There's plenty of room at the bottom. An invitation to open up a new field of physics.* "Engineering and Science", Vol. XXIII, No. 5: 22–36.

Fisher E. and Mahajan R.L. 2006. *Contradictory intent? US federal legislation on integrating societal concerns into nanotechnology research and development.* "Science and Public Policy", Vol. 33/1: 5–16.

Fisher E., Mahajan R., Mitcham C. 2006. *Midstream modulation of technology: Governance from within.* "Bulletin of Science, Technology & Society", Vol. 26(6): 485–496.

Fleischer T. 2002. *Technikfolgenabschätzungen zur Nanotechnologie – Inhaltliche und konzeptionelle Überlegungen.* „Theorie und Praxis", Vol. 11: 111–122.

Fleischer T. 2011. *Nanotechnologie*. In: Maring M. (Hrsg.), *Fallstudien zur Ethik in Wissenschaft, Wirtschaft, Technik und Gesellschaft*. Karlsruhe: KIT Scientific Publishing, 176–184.

Florida R. 1999. *The role of the university: Leveraging talent, not technology*. "Issues in Science and Technology", Summer; http://www.issues.org.

Fukuyama F. 2002. *Our Posthuman Future. Consequences of the Biotechnology Revolution*. London: Profile Books.

Funtowicz S.O. and Ravetz J.R. 1993. *Science for the post-normal age*. "Futures", Vol. 25: 739–755.

Funtowicz S.O. and Ravetz J.R. 1994. *Uncertainty, complexity and post-normal science*. "Environmental Toxicology and Chemistry", Vol. 13, No. 12: 1881–1885.

Funtowicz S.O. and Ravetz J.R. 2003. *Post-Normal Science*. In: International Society for Ecological Economics, Internet Encyclopaedia of Ecological Economics, 1–10.

Gammel S., Lösch A., Nordmann A. 2009. *Jenseits von Regulierung: Zum politischen Umgang mit der Nanotechnologie*. Heidelberg: Akademische Verlagsgesellschaft AKA.

Geertz C. 1973. *The Interpretation of Cultures: Selected Essays*. New York: Basic Books.

Gibbons M. 1999. *Science's new social contract with society*. "Nature", Vol. 402: C81-C84.

Gibbons M., Limoges C., Nowotny H., Schwartzman S., Scott P., Trow M. 1994. *The New Production of Knowledge: The Dynamics of Science and Research in Contemporary Societies*. London: Sage.

Gibson I., Rosen D.W., Stucker B. 2010. *Additive Manufacturing Technologies. Rapid Prototyping to Direct Digital Manufacturing*. New York et al.: Springer Science+Business Media.

Godin B. 2006. *The linear model of innovation: The historical construction of an analytical framework*. "Science, Technology & Human Values", Vol. 31: 639–667.

Gorman M.E., Groves J.F., Shrager J. 2004. *Societal dimensions of nanotechnology as a trading zone: Results from a pilot project*. In: Baird D., Nordmann A., Schummer J. (eds), *Discovering the Nanoscale*. Amsterdam: IOS Press, 63–73.

Grunwald A. 2002. *Technikfolgenabschätzung – eine Einführung*. Berlin: Edition Sigma.

Grunwald A. 2004. *Vision Assessment as a new element of the Technology Futures Analysis Toolbox.* In: Scapolo F. and Cahill E. (eds.), *New Horizons and Challenges for Future-oriented Technology Analysis*, Proceedings of the EU-US Scientific Seminar: New Technology Foresight, Forecasting & Assessment Methods, JRC-IPTS Seville, May 13–14 2004.

Grunwald A. 2008a. *Auf dem Weg in eine nanotechnologische Zukunft. Philosophisch-ethische Fragen.* Freiburg: Alber Verlag.

Grunwald A. 2008b. *Nanoparticles: Risk Management and the Precautionary Principle.* In: Jotterand F. (ed.), *Emerging conceptual, ethical and policy issues in bionanotechnology.* Berlin: Springer, 85–102.

Grunwald A. 2010. *From Speculative Nanoethics to Explorative Philosophy of Nanotechnology.* "Nanoethics", Vol. 4: 91–101.

Grunwald A. and Julliard Y. 2007. *Nanotechnology – steps towards understanding human beings as technology?* "NanoEthics", Vol. 1: 77–87.

Guston D. 2000. *Between Politics and Science: Assuring the Integrity and Productivity of Research.* Cambridge University Press.

Habermas J. 2001(2003). *Die Zukunft der menschlichen Natur. Auf dem Wege zu einer liberalen Eugenik?* Frankfurt/Main: Suhrkamp. [*The Future of Human Nature.* Cambridge: Polity Press].

Hacking I. 1983. *Representing and Intervening: Introductory Topics in the Philosophy of Natural Science.* Cambridge University Press.

Harremoes P., Gee D., MacGravin M., Stirling A., Keys J., Wynne B., Guedes Vaz S. 2002. *The precautionary principle in the 20$^{th}$ century. Late lessons from early warnings.* London: Sage.

Haslinger J., Hauser Ch., Hocke P., Fiedeler U. 2012. *Ein Teilerfolg der Nanowissenschaften? Eine Inhaltsanalyse zur Nanoberichterstattung in repräsentativen Medien Österreichs, Deutschlands und der Schweiz.* Wien: Institut für Technikfolgen-Abschätzung der Österreichischen Akademie der Wissenschaften.

Haum R., Petschow U., Steinfeldt M., Gleich A. von 2004. *Nanotechnology and Regulation within the Framework of the Precautionary Principle.* Schriften des IÖW 173/04, Berlin: Eigenverlag.

HLEG 2004. *Report of High Level Expert Group: Converging Technologies – Shaping the Future of European Societies.* Luxembourg: Office for Official Publications of the European Communities, EUR 21357.

Hook C.C. 2004. *The techno sapiens are coming.* "Christianity Today Magazine"; www.christianitytoday.com/ct/2004/001/1.36.html

Hughes J. 2006. *Human Enhancement and the Emergent Technopolitics of the 21st Century*. In: Bainbridge W.S. and Roco M.C. (eds.), *Managing Nano-Bio-Info-Cogno Innovations. Converging Technologies in Society*. Dordrecht: Springer, 285–309.

Jasanoff S. 2005. *Designs on Nature: Science and Democracy in Europe and the United States*. Princeton University Press.

Jasanoff S. (ed) 2004. *States of Knowledge: The Co-production of Science and The Social Order*. London: Routledge.

Joly P.B. and Kaufmann A. 2008. *Lost in translation? The need for 'upstream engagement' with nanotechnology on trial*. "Science as Culture", Vol. 17: 1–23.

Jonas H. 1979. *Das Prinzip Verantwortung*. Frankfurt/Main: Suhrkamp.

Joy B. 2000. *Why the future doesn't need us: Our most powerful 21st-century technologies – robotics, genetic engineering, and nanotech – are threatening to make humans an endangered species*. "Wired", Vol. 8 (4); www.wired.com/wired/archive/8.04/joy.html

Kaiser M., Kurath M., Maasen S., Rehmann-Sutter Ch. 2010. *Governing Future Technologies. Nanotechnology and the Rise of an Assessment Regime*, Springer.

Keiper A. 2007. *Nanoethics as a discipline?* "The New Atlantis. A Journal of Technology & Science", 55–67.

Khushf G. 2004. *The Ethics of Nanotechnology. Visions and Values for a new Generation of Science and Engineering*. In: National Academy of Engineering, *Emerging Technologies and Ethical Issues in Engineering*. Washington, 29–55.

Kjolberg K.L. and Wickson F. 2010. *Nano meets Macro. Social Perspectives on Nanoscale Sciences and Technologies*. Pan Stanford Publishing.

Knorr Cetina K. 1995. *Laboratory studies: The cultural approach to the study of science*. In: Jasanoff S., Markle G., Petersen J. C., Pinch, T. (eds), *Handbook of Science and Technology Studies*. Thousand Oaks: SAGE Publication, 140–166.

Knorr Cetina K. 1999. *Epistemic Cultures: How the Sciences Make Knowledge*. Cambridge: Harvard University Press.

Krimsky S. 2003. *Science in the private interest: Has the lure of profits corrupted biomedical research?* Lanham: Rowman & Littlefield.

Latour B. 1987. *Science in Action: How to Follow Scientists and Engineers Through Society*. Milton Keynes, Philadelphia: Open University Press.

Latour B. 1993. *We Have Never Been Modern*. Cambridge: Harvard University Press.

Latour B. and Woolgar S. 1979. *Laboratory Life: The Social Construction of Scientific Facts*. Beverly Hills, London: Sage Publications.

Lewenstein B. 2005. *What counts as a 'social and ethical issue' in nanotechnology?* "HYLE – International Journal for Philosophy of Chemistry", Vol. 11: 5–18.

MacDonald C. 2004. *Nanotech is Novel;the Ethical Issues Are Not.* "The Scientist", Vol.18:3.

Macnaghten P.M., Kearnes M.P., Wynne B. 2005. *Nanotechnology, governance, and public deliberation: What role for the social sciences?* "Science Communication", Vol. 27: 1–24.

Maienschein J. 2002. *Innocent reflections on science and technology policy.* "Technology in Society", Vol. 24: 133–143.

Mali F. 2009. *Bringing Converging Technologies Closer to Civil Society. The Role of Precautionary Principle in Risk Technology Assessment.* "Innovation: The European Journal of Social Science Research", Vol. 22, No 1: 53–75.

Merton R. 1973. *Sociology of Science.* Chicago University Press.

Miller S.E. 2007. *Converging technologies: Innovation, legal risks, and society.* In: Roco M.C. and Bainbridge W.S., *Nanotechnology: Societal Implications II. Individual Perspectives.* Dordrecht: Springer, 158–161.

Mnyusiwalla A., Daar A.S., Singer P.A. 2003. *Mind the gap: Science and ethics in nanotechnology.* "Nanotechnology", Vol. 14: R9–R13.

Mody C. 2004. *How probe microscopists became nanotechnologists.* In Baird D., Nordmann A., Schummer J. (eds), *Discovering the Nanoscale.* Amsterdam: IOS Press, 119–134.

Moor J. and Weckert J. 2004. *Nanoethics: assessing the nanoscale from an ethical point of view.* In: Baird D., Nordmann A., Schummer J. (eds), *Discovering the nanoscale*, Amsterdam: IOS Press, 301–310.

Müller O. 2010. *Zwischen Mensch und Maschine.* Frankfurt/Main: Suhrkamp.

National Research Council 2007. *The National Science Foundation's Materials Research Science and Engineering Centers Program: Looking Back, Moving Forward.* Report by the MRSEC Impact Assessment Committee, Solid State Sciences Committee, National Research Council.

National Science and Technology Council 1999. *Nanotechnology – Shaping the World Atom by Atom.* Washington: US Government.

National Science and Technology Council 2000. *The National Nanotechnology Initiative: The Initiative and its Implementation Plan.* Committee on

Technology, Subcommittee on Nanoscale Science, Engineering and Technology. Washington: U. S. Government Printing Office.

NEST 2004. *Interdisciplinarity in Research*, Paper by the European Union Research Advisory Board. April; http://ec.europa.eu/research/eurab/pdf/eurab_04_009_interdisciplinarity_research_final.pdf

Nordmann A. 2004. *Nanotechnology's worldview: New space for old cosmologies.* "Technology and Society Magazine", IEEE, Vol. 23: 48–54.

Nordmann A. 2007. *If and Then: a critique of speculative nanoethics.* "Nanoethics", Vol. 1: 31–46.

Nordmann A. and Rip A. 2009. *Mind the gap revisited.* "Natural Nanotechnology", Vol. 4: 273–274.

OECD 2011. *Science, Technology and Industry Scoreboard*; http://www.oecdilibrary.org/science-and-technology/oecd-science-technology-and-industry-scoreboard-2011_sti_scoreboard-2011-en

OECD Series on the Safety of Manufactured Nanomaterials: http://www.oecd.org/env/ehs/nanosafety/publicationsintheseriesonthesafetyofmanufacturednanomaterials.htm

Parker L. 1997. *The Engineering Research Centers (ERC) Program: An Assessment of Benefits and Outcomes*, Report from Engineering Education and Centers Division, Directorate for Engineering. Arlington-Virginia: National Science Foundation.

Paschen H., Coenen C., Fleischer T., Grünwald R., Oertel D., Revermann Ch. 2003. *TA-Projekt Nanotechnologie. Endbericht.* Arbeitsbericht Nanotechnologie Nr. 92, Berlin: Büro für Technikfolgen beim Deutschen Bundestag.

Pitt J.C. 2004. *The epistemology of the very small.* In: Baird D., Nordmann A., Schummer J., (eds), *Discovering the Nanoscale.* Amsterdam: IOS Press, 157–163.

Pitt J.C. 2005. *When is an image not an image?* "Techné: Research in Philosophy and Technology", Vol. 8(3): 23–33.

Pool R. 1997. *Beyond Engineering: How society shapes technology.* Oxford University Press.

Porter A.L. and Youtie J. 2009a. *How interdisciplinary is nanotechnology?* "Journal of Nanoparticle Research", Vol. 11(5): 1023–1041.

Porter A.L. and Youtie J. 2009b. *Where does nanotechnology belong in the map of science?* "Nature-Nanotechnology", Vol. 4: 534–536.

President's Council of Advisors on Science and Technology 2008. *The national nanotechnology initiative: Second assessment and recommendations of the national nanotechnology advisory panel*, Washington DC.

Rader M., Coenen Ch., Fleischer T., Luber B.-J., Quendt Ch. 2007. *Current trends in RTD policy on Converging Technologies*, Deliverable of the CONTECS project.

Rapp F. 1999. *Normative Technikbewertung. Erfahrungen mit der VDI-Richtlinie 3780*. Berlin: Sigma.

Riediker M. and Katalagarianakis G. 2010. *Compendium of Projects in the European Nano Safety Cluster*. Lausanne: Institute for Work and Health.

Roache R. 2008. *Ethics, speculation, and values*. "Nanoethics", Vol. 2: 317–327.

Roco M. C. 2004. *The US National Nanotechnology Initiative after 3 years (2001–2003)*. "Journal of Nanoparticle Research", Vol. 6: 1–10.

Roco M. and Bainbridge W. (eds.) 2001. *Societal Implications of Nanoscience and Nanotechnology*, Final Report from the Workshop at the National Science Foundation, Arlington in September 28–29, 2000, Dordrecht: Kluwer. http://www.wtec.org/loyola/nano/NSET.Societal.Implications/nanosi.pdf

Roco M.C. and Bainbridge W.S. (eds.) 2003. *Converging technologies for improving human performance*. Dordrecht: Kluwer.

Roco, M.C. and Bainbridge W.S. (eds.) 2005. *Nanotechnology: Societal Implications – Maximizing Benefit for Humanity*, Nanoscale Science, Engineering and Technology Subcommittee of the US National Science and Technology Council.

Roco M.C. and Bainbridge W.S. 2007. *Nanotechnology: Societal Implications II. Individual Perspectives*. Dordrecht: Springer.

Roco M.C., Bainbridge W.S., Tonn B., Whitesides G. (eds.) 2013. *Convergence of Knowledge, Technology and Society. Beyond Convergence of Nano-Bio-Info-Cognitive Technologies*. Heidelberg et al.: Springer.

Royal Society and The Royal Academy of Engineering (RS/RAE) 2004. *Nanoscience and nanotechnologies: Oppoortunities and uncertainties*. London: Royal Society.

Schmid G., Decker M., Ernst H., Fuchs H., Grünwald W., Grunwald A., Hofmann H., Mayor M., Rathgeber W., Simon U., Wyrwa D. 2003. *Small Dimensions and Material Properties. A Definition of Nanotechnology*. Europäische Akademie zur Erforschung von Folgen wissenschaftlich-technischer Entwicklung GmBH, Bad Neuenahr-Ahrweiler: Graue Reihe Vol. 35.

Schummer J. 2004. *Multidisciplinarity, interdisciplinarity, and patterns of research collaboration in nanoscience and nanotechnology.* "Scientometrics", Vol. 59: 425–465.

Shapira P., Youtie J., Porter A.L. 2010. *The emergence of social science research on nanotechnology.* "Scientometrics", Vol. 85: 595–611.

Slaughter S. and Leslie L. L. 1997. *Academic Capitalism. Politics, Policies, and the Entrepreneurial University.* Baltimore, London: John Hopkins University Press.

Smalley R.E. 2001. *Of chemistry, love and nanobots.* "Scientific America", Vol. 285(3): 76–77.

Snow C.P. 1961. *The Two Cultures and the Scientific Revolution.* New York: Cambridge University Press.

Stokes D.E. 1997. *Pasteur's Quadrant: Basic Science and Technological Innovation.* Washington: Brookings Institution Press.

Strand R. and Birkeland T. 2010. *The Science and Politics of Nano Images.* In: Kjolberg K.L., Wickson F., *Nano meets Macro. Social Perspectives on Nanoscale Sciences and Technologies.* Pan Stanford Publishing, 85–107.

TAB (Office of Technology Assessment at the German Parliament) 2008. *Konvergierende Technologien und Wissenschaften. Der Stand der Debatte und politischen Aktivitäten zu Converging Technologies.* Berlin: TAB Background Paper 16.

Tickner J., Raffensperger C., Myers N. 2004. *The Precautionary Principle in Action: A handbook.* Lowell: Science and Environment Health Network.

UNESCO 2003. *Universal Declaration on the Human Genome and Human Rights: from theory to practice*; http://unesdoc.unesco.org/images/0012/001229/122990eo.pdf

Wang J. 2013. *Nanomachines.* New York: John Wiley & Sons.

Whiteside K. 2006. *Precautionary Politics. Principle and Practice in Confronting Environmental Risk.* Cambridge, Massachusetts: The MIT Press.

Ziman J. 1978. *Reliable Knowledge.* Cambridge UK: Cambridge University Press.

Ziman J. 1998. *Why must scientists become more ethically sensitive than they used to be?* "Science", Vol. 282: 1813–1814.

Ziman J. 2000. *Real Science. What It Is, and What It Means.* Cambridge University Press.

# Index

**A**
Abriszewski, Krzysztof 60
Afeltowicz, Łukasz 9
Alhakami, Ali 150
Allhoff, Fritz 99, 101, 122, 143, 149-151, 153, 160
Altmann, Jürgen 128, 157
Amato, Ivan 138
Andler, Daniel (et al.) 174, 175-177, 186, 187-188
Andorno, Roberto 52-54
Apel, Karl-Otto 56

**B**
Bachmann-Medick, Doris 110
Bainbridge, William S. 79, 101, 111, 117, 135, 137-141, 153, 159, 166-167, 170-174, 177, 178-183, 186
Baird, Davis 177-178
Ball, Philip 156
Barthe, Yannick 33
Barry, Andrew 109, 112
Baum, Rudy 138, 140
Bauman, Zygmunt 19
Baumgartner, Christoph 160
Beck, Silke 37, 44, 46-47
Beck, Ulrich 1-2, 5-6, 13, 15-23, 27-28, 32-34, 36-39, 50-51, 53, 62-63, 65
Bennett, Ira 116
Bensaude-Vincent, Bernadette 107
Bessette, Joseph 37
Bhagwati, Jagdish 4
Bhushan, Bharat 90-91
Bijker, Wiebe E. 51
Binks, Peter 139
Binnig, Gerd 106

Bińczyk, Ewa vii-ix, 1, 47, 55, 143
Birkeland, Tore 113, 115-116
Birnbacher, Dieter 56
Bouchard, Raymond 174
Böhme, Gernot 12, 168
Braun, Kathrin 47
Brundtland, Gro Harlem 59
Brune, Harald (et al.) 81, 83-93, 118, 120-121, 123-127, 129, 131-132, 134, 136-137, 139-142, 150, 152, 154, 156-158, 160
Bucchi, Massimiano 8, 11-12, 14, 168, 170
Bush, Vannevar 99, 114

**C**
Callon, Michel 33
Carnap, Rudolf 86
Chomsky, Noam 6
Christensen, Clayton M. 139, 178
Collingridge, David viii, 164
Collins, Harry 36
Colvin, Vicki 126, 144-145, 187
Crichton, Michael 139
Cunningham-Burley, Sarah 41

**D**
Dewey, John 47
Donaldson, Ken 145-146, 186
Drexler, Eric 105-106, 136-137, 138, 140
Durant, Darrin 35-36, 41
Dybel, Paweł 19, 21

**E**
Eigler, Donald M. 81, 106, 115
Epstein, Steven 41
Etzkowitz, Henry 105

**F**
Feynman, Richard P. 79-84, 88, 90-91, 105-106, 114, 136, 178, 189
Fisher, Erik 108, 111
Fishkin, James 37
Fleischer, Torsten 160, 184-185, 186
Florida, Richard 173
Frege, Gottlob 114
Fukuyama, Francis 66, 168
Fuller, Steve 11, 35, 44, 50
Funtowicz, Silvio O. 78, 94-97

**G**
Gammel, Stefan 87
Geertz, Clifford 110
Gibbons, Michael 13-14, 97, 104-105, 109
Gibson, Ian 83, 167
Giddens, Anthony 15, 21, 32, 34, 36, 39, 60, 62
Godin, Benoît 3, 9, 11, 104
Gorman, Michael E. 111
Grunwald, Arnim 87, 141, 157, 160-166, 186
Guston, David H. 99, 104, 109

**H**
Habermas, Jürgen 159, 168
Hacking, Ian 110
Hanson, Robin 137
Harding, Sandra 62
Harremoes, Poul 124-125, 130-131
Haslinger, Julia 186
Haum, Rüdiger 128, 132-133
Held, David 37
Helmholtz, Hermann von 86
Hempel, Carl Gustav 86
Hertz, Heinrich 91
Hook, Christopher 159
Hughes, James J. 168

**I**
Ihde, Don 60
Irwin, Alan 35-36

**J**
Jasanoff, Sheila 40, 109-110
Joly, Pierre-Benoit 112
Jonas, Hans 57-59, 129, 132, 162
Jones, Mark Peter 3, 9
Joy, Bill 136-137, 139
Julliard, Yannick 165

**K**
Kaiser, Mario 101, 186
Kant, Immanuel 58
Katalagarianakis, Georgios 80, 147-149, 186
Kaufmann, Alain 112
Keiper, Adam 101, 161
Kerr, Anne 41
Khushf, George 94, 152, 159, 160
Kjolberg, Kamilla Lein 77-81, 88, 97-101, 103, 105-112, 186
Klaassen, Johann A. 52-53
Klein, Naomi 5
Knorr Cetina, Karin 110
Krimsky, Sheldon 8, 10, 49-50, 173
Kropp, Cordula 47
Król, Marcin 37
Kuhn, Thomas 94
Kurczewska, Joanna 33
Kusch, Martin 40
Kutyła, Julian 20

**L**
Laird, Frank N. 42-44
Lash, Scott 21, 32, 34, 39, 62
Lascoumes, Pierre 33
Latour, Bruno 1, 4, 6-9, 20, 22-31, 33, 38-39, 50-51, 53-54, 56, 62-63, 99, 102, 110
Lave, Rebecca 10, 12

Lengwiler, Martin 34-35, 45
Lente, Harro van 35-36, 48
Leslie, Larry 105
Levitas, Ruth 32-33
Levy, David J. 57
Lewenstein, Bruce V. 111
Leydesdorff, Loet 105
Lin, Patrick 99, 101, 122, 143, 149-151, 153, 160
Lövbrand, Eva 37, 44, 46-47
Luhmann, Niklas 16, 62-63

**M**
MacDonald, Chris 160
MacIntyre, Alasdair 63
Macnaghten, Phil 109, 111
Mahajan, Roop L. 108
Maienschein, Jane 111
Mali, Franc 168-170
Marres, Noortje 36, 47
Merton, Robert 11, 97, 170
Miller, Sonia E. 186-187
Mills, Charles Wright 39
Mirowski, Philip 3, 10, 12
Mnyusiwalla, Anisa 87, 101, 116-117, 146, 152, 155, 163
Mody, Cyrus C.M. 106
Moor, James 157, 159, 160, 162
Mucha, Janusz 8, 11
Müller, Oliver 167

**N**
Nahuis, Roel 35-36, 48
Nordmann, Alfred 114, 161-164
Nowotny, Helga 13

**O**
Okrasiński, Krzysztof 46
Ostolski, Adam 20

**P**
Papadopoulos, Dimitris 37
Parker, Linda 107

Paschen, Herbert 145
Passmore, John 22
Perelman, Michael 54
Pielke, Roger 37, 44, 46-47
Pitt, Joseph C. 113, 115
Pool, Robert 177
Porter, Alan L. 117
Postman, Neil 49

**R**
Rader, Michael 175
Raffensperger, Carolyn 124, 130, 134
Randalls, Samuel 10, 12
Rapp, Friedrich 154
Ravetz, Jerome R. 78, 94-97
Rhoades, Gary 9-10
Riediker, Michael 80, 147-149, 186
Rip, Arie 161-164
Roache, Rebecca 161, 166
Roco, Mihail C. 79, 101, 108, 111, 117, 135, 137-141, 153, 159, 166-167, 170-174, 175, 177, 178-183, 186, 192
Rohrer, Heinrich 106
Rowland, Nicholas J. 23
Roystor, Ivor 9

**S**
Sarewitz, Daniel 116
Schettler, Ted 124, 130, 134
Schmid, Günter 93-94, 183-184
Schummer, Joachim 117
Schweizer E.K. 106
Scott, Peter 13
Selinger, Evan 60
Shapira, Philip 116-117
Sismondo, Sergio 9, 11, 46
Slaughter, Sheila 9-10, 105
Sloterdijk, Peter 20
Slovic, Paul 150
Smalley, Rick 92, 138, 140
Sombart, Werner 49

Soros, George 20
Stankiewicz, Piotr 26, 41, 45, 47, 55
Stern, Nicholas 20-21
Stępień, Tomasz vii-ix, 77
Stiglitz, Joseph E. 4
Stokes, Donald E. 104
Strand, Roger 113, 115-116
Strydom, Piet 32, 56-57, 63
Sunstein, Cass R. 39-40
Sutowski, Michał 20
Szahaj, Andrzej 3

**T**
Taniguchi, Norio 184
Tickner, Joel A. 124, 130-131, 134, 177
Turner, Stephen 44
Tutton, Richard 41
Twardowski, Tomasz 45

**V**
Van Horn, Robert 3, 10, 12

**W**
Waals, Johannes Diderik van der 83
Wallerstein, Immanuel 1, 5-6
Wang, Joseph 167
Wasilewska, Agnieszka 46
Weber, Max 99
Weckert, John 157, 159, 160, 162
Welsh, Rick 54
Whiteside, Kerry H. 168
Wickson, Fern 77-81, 88, 97-101, 103, 105-112, 186
Woolgar, Steve 110
Wróbel, Szymon 19, 21
Wynne, Brian 35-36, 40

**Y**
Yearley, Steven 11, 35-36, 47
Youtie, Jan 117

**Z**
Zacher, Lech W. 51, 65
Ziman, John 8, 97-98, 104-105, 170, 180
Zybertowicz, Andrzej 15, 17

**Comparative Studies on Education, Culture and Technology**
**Vergleichende Studien zur Bildung, Kultur und Technik**

Edited by / Herausgegeben von
Tomasz Stępień

Vol. / Bd. 1   Tomasz Stępień / Annette Deschner / Mojca Kompara / Adriana Merta-Staszczak: Spatialisation of Education. Migrating Languages – Cultural Encounters – Technological Turn. 2013.

Vol. / Bd. 2   Anton Hilckman: Gesammelte Werke. Schriften zur philosophischen Pädagogik Teil 1. Bildung – Begeisterung – Freiheit. Bearbeitet, kommentiert und herausgegeben von Tomasz Stępień. 2014.

Vol. / Bd. 3   Anton Hilckman: Gesammelte Werke. Schriften zur philosophischen Pädagogik Teil 2. Christliche Philosophie. Bearbeitet, kommentiert und herausgegeben von Tomasz Stępień. 2014.

Vol. / Bd. 4   Ewa Bińczyk / Tomasz Stępień: Modeling Technoscience and Nanotechnology Assessment. Perspectives and Dilemmas. 2014.

www.peterlang.com

 www.ingramcontent.com/pod-product-compliance
Ingram Content Group UK Ltd.
Pitfield, Milton Keynes, MK11 3LW, UK
UKHW041912140426
5217IPUK00002B/15